# Springer Series in Computational Physics

Editors:
W. Beiglböck,   H. Cabannes,
H. B. Keller,   J. Killeen,   S. A. Orszag

François Thomasset

# Implementation of Finite Element Methods for Navier-Stokes Equations

With 86 Illustrations

Springer-Verlag
New York Heidelberg Berlin

Dr. François Thomasset
Domaine de Voluceau
Rocquencourt
B.P. 105
F-78150 Le Chesnay
France

Editors:

Wolf Beiglböck

Institut für Angewandte Mathematik
Universität Heidelberg
Im Neuenheimer Feld 5
D-6900 Heidelberg 1
Federal Republic of Germany

Henri Cabannes

Mécanique Théoretique
Université Pierre et Marie Curie
Tour 66. 4, Place Jussieu
F-75005 Paris
France

Stephen A. Orszag

Department of Mathematics
Massachusetts Institute of Technology
Cambridge, Massachusetts 02139
USA

H. B. Keller

Applied Mathematics 101-50
Firestone Laboratory
California Institute of Technology
Pasadena, California 91125
USA

John Killeen

Lawrence Livermore Laboratory
P.O. Box 808
Livermore, California 94551
USA

Library of Congress Cataloging in Publication Data
Thomasset, François.
    Implementation of finite element for Navier-
Stokes equations.

    (Springer series in computational physics)
    1. Fluid dynamics.   2. Navier-Stokes equations—
Numerical solutions.   3. Finite element method.   I. Title.
II. Series.
QA911.T46         515.3′53         81-9169
                                    AACR2

9 8 7 6 5 4 3 2 1

ISBN 978-3-642-87049-1         ISBN 978-3-642-87047-7 (eBook)
DOI 10.1007/978-3-642-87047-7

# Contents

# Introduction

In structure mechanics analysis, finite element methods are now well established and well documented techniques; their advantage lies in a higher flexibility, in particular for:

(i) The representation of arbitrary complicated boundaries;
(ii) Systematic rules for the developments of stable numerical schemes approximating mathematically wellposed problems, with various types of boundary conditions.

On the other hand, compared to finite difference methods, this flexibility is paid by: an increased programming complexity; additional storage requirement.

The application of finite element methods to fluid mechanics has been lagging behind and is relatively recent for several types of reasons:

(i) Historical reasons: the early methods were invented by engineers for the analysis of torsion, flexion deformation of bearns, plates, shells, etc... (see the historics in Strang and Fix (1972) or Zienckiewicz (1977)).
(ii) Technical reasons: fluid flow problems present specific difficulties: strong gradients,[1] of the velocity or temperature for instance, may occur which a finite mesh is unable to properly represent; a remedy lies in the various upwind finite element schemes which recently turned up, and which are reviewed in chapter 2 (yet their effect is just as controversial as in finite differences).

Next, waves can propagate (e.g. in ocean dynamics with shallowwaters equations) which will be falsely distorted by a finite non regular mesh, as Kreiss (1979) pointed out.

We are concerned in this course with the approximation of incompressible, viscous, Newtonian fluids, i.e. governed by Navier Stokes equations.

We leave aside many problems of practical interest which have been solved by finite element techniques:

For transonic flows with shocks, see Bristeau (1977, 1978), Glowinski and Pironneau (1976) and the references of these papers.

---

[1] Shocks will not be considered in this course.

Viscous non Newtonian fluids are studied by Nickell, Tanner and Caswell (1974), Engelman (1979) (with application to blood flow past prosthetic valves), Zienckiewicz and Godbole (1975).

Chavent (1979), Jaffre (1979b) considered flows in porous media with applications to oil production.

Moving meshes (to deal with free surfaces or multi fluid problems) have been seldom used: see Hughes, Liu and Zimmermann (1978, generalized ALE technique). An alternative technique using a fixed mesh for the simulation of two fluid flows is proposed by Dervieux and Thomasset (1979).

Further applications can be found e.g. in Zienckiewicz (1977).

The plan of the course is as follows: in the first chapter some useful finite element methods are presented. Next in chapter 2 the upwinding techniques in finite elements are reviewed and in chapter 3 the various formulations available to deal with the incompressibility conditions are presented. In chapter 4, the techniques of previous chapters are combined for the approximation of Navier Stokes equations; some numerical results are presented, although it is felt that further tests should be necessary, in order to allow definite assessments about the performance of the various methods. Finally chapter 5 surveys the different techniques related to automatic mesh generation, and solution of systems of linear or non linear equations.

Unless the contrary is specified[2] the methods presented in this paper extend to three dimensional problems without other difficulties, than programming, solving very large systems of equations, and interpreting the results (which by the way are not small difficulties). The actual challenge is the application of finite element methods to large 3D problems will require the development of parallel processors; Pironneau (1979) suggested some methods for the solution of Navier Stokes equations using parallel processors with SIMD architecture.

For the theory of finite element methods, the reader is referred, e.g. to Ciarlet (1978), Babuška and Aziz (1972), Strang and Fix (1972), Oden and Reddy (1976b), Girault and Raviart (1979), de Boor (1974), Miller (1978), Raviart (1979); for mathematical problems concerning partial differential equations, see for instance Agmon (1965), Lions (1969, 1977), Nečas (1967), Temam (1977).

Finite element methods offer a great flexibility to solve general partial differential equations; on the other hand, the preliminary investments in programmation are important, and cannot be efficiently developed by isolated programmers. Thus in order to remedy the development of redundant programs by various institutions and industries. I.R.I.A. created the club MODULEF[3] about five years ago. The goal of this club is to offer a structure where the members, from industry and university, can meet and share their

---

[2] e.g. the methods with the stream function are 2-dimensional.

[3] Begis and Perronnet (1979).

experience and packages; libraries of routines are already developed concerning: mesh generation; solvers of linear systems; analysis of structural mechanics problems; a few modules (at this date) concern fluid mechanics (Navier Stokes equations). Further effort should be done for easier learning facilities and readability of Fortran programs.

For the realization of this course I am gratefully indebted to a number of persons: first of all to my friends at I.N.R.I.A. for helpful discussions and supporting fellowship, specially to A. Dervieux, J. F. Bourgat and M. O. Bristeau who supplied some figures; to J. Periaux from Avions-Marcel-Dassault/Bréguet-Aviation who supplied a numerical example in an industrial configuration; to my friends of Departement d'Analyse Numérique of Universités Paris 6 and Paris 11; to B. L. Hua from L.O.P./Muséum, for helpful comments; to Mme Kurinckx-Longrée for the efficient and beautiful typing of the manuscript; to my wife Annic for her patient support during these last few months.

Thanks are also due to the Von Karman Institute for Fluid Dynamics (Belgium) where this course was displayed (V.K.I. Lecture series, Computational Fluid Dynamics, March 26–29, 1980. Finally am gratefully indebted to Pr. R. Temam, to whom I owe my initiation to Numerical Analysis and who introduced me to V.K.I., and to Pr. R. Glowinski and Pr. P. A. Raviart for their encouraging and stimulating discussions.

# Notations

$\Omega$ = physical (bounded) domain of calculation

$\Gamma = \partial\Omega$ = boundary of $\Omega$

$\Gamma_0, \Gamma_1, \Gamma_2 \ldots$ = parts of $\Gamma$ (in the case of the flow around an obstacle, $\Gamma_0$ is the exterior boundary)

$t$ = time variable

$x = \{x_1, x_2\}$ or $\{x_1, x_2, x_3\}$ = generic point in $\Omega$

$dx = dx_1 dx_2$ or $dx_1 dx_2 dx_3$ = infinitesimal area

$ds$ = infinitesimal length on $\Gamma$

$L^2(\Omega)$ = set of square integrable functions
$$= \{v: \int_\Omega |v(x)|^2 \, dx < +\infty\}$$

$H^1(\Omega) = \{v \in L^2(\Omega) = \int_\Omega |\text{grad } v|^2 \, dx < +\infty\}$

$H_0^1(\Omega)$ = set of functions in $H^1(\Omega)$ vanishing on the boundary
$$= \{v \in H^1(\Omega): v|_\Gamma = 0\}$$

$H(\text{div}, \Omega) = \{\mathbf{v} = \{v_1, v_2\}: v_i \in L^2(\Omega), \text{div } v \in L^2(\Omega)\}$

$H_0(\text{div}, \Omega) = \{\mathbf{v} \in H(\text{div}, \Omega): \text{div}\mathbf{v} = 0 \text{ in } \Omega, \mathbf{v} \cdot \mathbf{n} = 0 \text{ on } \Gamma\}$

$\mathbf{n} = \mathbf{n}(x)$ = unit vector normal to the boundary at a point $x$, pointing outwards of the domain.

$u, \theta$ = solutions of model problems: Poisson's equation and advection-diffusion equation

$\mathbf{u}$ = velocity field ($u$ in 1 dimension)

$p$ = pressure

$\sigma_{ij}$ = stress tensor

$\mathbf{f} = \{f_i\}$ = body force

$\kappa$ = diffusivity coefficient

$\kappa_{ij}$ = diffusivity tensor

$\nu$ = kinematic viscosity

$\psi$ = stream function

$\omega$ = vorticity (or a relaxation parameter, according to the context).

$\lambda$ = value of $p$ (or $\omega$) on the boundary.

$v, \mathbf{v}, q$ = trial functions

$\mathcal{T}_h$ = triangulation (or quadrangulation) with meshsize $h$

$NS$ = number of vertices

$NT$ = number of elements (triangles, quadrangles or other)

$N$ = number of (discrete) equation

$h$ or $x$ = meshsize

$u_h, \theta_h, \mathbf{u}_h, p_h, \psi_h, \omega_h$ = finite element approximations to $u, \theta, \mathbf{u}, p, \psi, \omega$.

$T$ = an element (generally a triangle)

$K$ = an element (generally a quadrangle)

$\lambda_i, i = 1, 2, 3$ = barycentric coordinates (or area coordinates) on a triangle $T$

$\hat{T}$ = the reference triangle

$\hat{K}$ = the reference quadrangle

$\xi = \{\xi_1, \xi_2\}$ = generic point in the reference element $\hat{T}$ or $\hat{K}$.

$F_T$ or $F_k$ = transformation from $\hat{T}$ or $\hat{K}$, to $T$ or $K$.

$J_T$ or $J_K$ = jacobian $= \left| \dfrac{\partial F_T}{\partial \xi} \right|$ or $\left| \dfrac{\partial F_K}{\partial \xi} \right|$

$\partial T$ = boundary of $T$

$\partial K$ = boundary of $K$

$\mathbf{n}_T(x)$ = unit vector normal to $\partial T$ at a point $s$, pointing outwards of $T$ (or $\mathbf{n}$ if this is not ambiguous)

$\mathbf{n}_K$: as above

$P_k$ ($k = 0, 1, 2, \ldots$) = set of polynomials in 2 variables of total degree $\leq k$

$Q_k$ ($k = 0, 1, 2, \ldots$) = set of polynomials in 2 variables, of degree $\leq k$ with respect to *each* variable

$\alpha$ = upwinding parameter

$\varepsilon$ = penalization parameter

$\rho$ = parameter of Uzawa's algorithm

$\phi_i, \phi_p$ = basis functions

$w_m$ = scalar basis functions, $P_1$ non conforming element

$\mathbf{w}_a, \mathbf{w}_m$ = vector valued, divergence free basis functions ($P_1$ non conforming element)

$\left. \begin{array}{l} a(u, v) \\ b(p, v) \end{array} \right\}$ bilinear forms

$J(v)$ = cost functional in a minimization process

$\mathcal{L}(\cdot, \cdot)$ = lagrangian functional

# 1. Elliptic Equations of Order 2: Some Standard Finite Element Methods

## 1.1. A 1-Dimensional Model Problem: The Basic Notions

Consider the following equation, to be satisfied on the real interval $0<x<1$:

$$Lu \equiv -\frac{d}{dx}\kappa(x)\frac{du}{dx} + \mu(x)u = f(x) \tag{1}$$

with the boundary conditions:

$$u(0)=0 \qquad (Dirichlet^4 \text{ boundary condition}) \tag{2}$$

$$\frac{du}{dx}(1)=0 \qquad (Neumann^5 \text{ boundary condition}). \tag{3}$$

The key of finite element methods (as any galerkin method) is to require for the residual $(Lu-f)$ to be orthogonal to all $v$ lying in a proper vector space:

$$(Lu-f, v)=0, \qquad \forall\, v \in V,^6 \tag{4}$$

where the parenthesis $(\cdot, \cdot)$ stand for the $L2$ scalar product over the domain of integration:

$$(u, v) = \int_{\Omega} uv\, dx, \qquad \Omega = \,]0, 1[$$

(this classical notation will be used throughout the course).

Equation (4) needs to be transformed, for several reasons, the first of which is that it contains second order derivatives which are difficult to approximate[7] in the framework of finite element methods. Therefore let us perform an integration by parts in (4):

$$\left(\kappa\frac{du}{dx}, \frac{dv}{dx}\right) + (\mu u, v) = (f, v) + \kappa\frac{du}{dx}v\bigg|_1 - \kappa\frac{du}{dx}v\bigg|_0 \tag{5}$$

Using the natural boundary condition (3), the first boundary term (at $x=1$) cancels out. The second term (at $x=0$) cancels if $v$ satisfies the same

---

[4] or "essential" boundary condition.
[5] "natural" boundary condition.
[6] $\forall\, v \in V$ means: "for all $v$ lying in $V$."
[7] such as it is.

homogeneous boundary condition as $u$:

$$v(0)=0$$

Thus we come to the following formulation of the original problem (1): Find $u$ such that:

$$u(0)=0$$
$$a(u,v)=(f,v), \qquad \forall \, v \text{ such that } v(0)=0 \tag{6}$$

with the notation:

$$\boxed{a(u,v)=\int_0^1 \left( \kappa \frac{du}{dx} \frac{dv}{dx} + \mu u v \right) dx.} \tag{7}$$

Note the *symmetric* role played by $u$ and $v$ in this formulation. In fact we shall take $v$ in the *same* class as $u$, which remains to define.

Before clearing this point, we note that (6) may be formulated as an optimization problem (of which (6) is the Euler equation): Find $u$ such that $u(0)=0$, and:

$$J(u)= \min_{v \mid v(0)=0} J(v)$$

with

$$J(v)=\tfrac{1}{2}a(v,v)-(f,v)$$

Over which space shall we minimize $J(v)$?

The integrals involved in (6) and (7) must be finite, and the class should be a complete Hilbert space, so that the minimum can be found in this class. This leads to the definition:
$V$= set of all $v$ such that:

$$\int_0^1 |v|^2 \, dx < +\infty$$

$$\int_0^1 \left| \frac{dv}{dx} \right|^2 dx < +\infty \tag{8}$$

$$v(0)=0$$

(we assume[8] that $\kappa(x)$ is bounded away from 0:

$$\kappa(x) \geq 0, \qquad \kappa(x) \geq \kappa_0 > 0.\,[9]$$

Note that the Dirichlet boundary condition is included in the definition (8) of

---

[8] This excludes the case of cylindrical coordinates, where $\kappa(x)=0(1/x)$ near $x=0$. For a study of this case, see Crouzeix and Thomas (1973).

[9] Then $\sqrt{a(v,v)}$ defines a *norm* on $V$, equivalent to the usual Sobolev norm:

$$\sqrt{\int_0^1 \left( |v|^2 + \left| \frac{dv}{dx} \right|^2 \right) dx}.$$

space $V$ ($v(0)=0$), but *not* the Neumann condition. Indeed the first boundary condition does make sense for functions satisfying (8), while the second (Neumann) condition does not.

Finally we get the following problem:

Find $u \in V$ such that:

$$a(u,v)=(f,v), \qquad \forall\, v \in V \tag{9}$$

or, equivalently:

$$\underset{v \in V}{\text{Min}}\, J(v)=\tfrac{1}{2}a(v,v)-(f,v) \tag{10}$$

Problem (9) is often called the "*variational* formulation" of differential equation (1). However we have in view the solution of Navier Stokes equations where no simple variational principle can be involved. Therefore the term "*weak formulation*" seems better suited for our purpose.

Finite element methods work on the weak formulation: they involve the construction of a finite dimensional subspace of $V$:

$$V_h \subset V$$

and require that (9) should be true for members of $V_h$ instead of $V$.

Find $u_h \in V_h$ such that

$$a(u_h, v_h)=(f, v_h), \qquad \forall\, v_h \in V_h \tag{11}$$

Let us describe the simplest of finite elements: the 1D-piecewise *linear* element.

We give on $)0, 1($ a subdivision:

$$0 = x_0 < x_1 < x_2 < \cdots < x_N = 1.$$

The functions $v_h \in V_h$ are required to satisfy the following requisites:

$v_h$ is continuous over $(0, 1)$;
$v_h$ is piecewise linear per interval;
$v_h(0) = 0$

Then it is easily seen that $V_h$ is generated by the following set of *basis*[10] functions ("roof" functions Figure 1)

$$x \le x_{i-1}: \phi_i(x)=0$$

$$x_{i-1} \le x \le x_i \quad : \phi_i(x)=\frac{x-x_{i-1}}{x_i-x_{i-1}}$$

$$x_i \quad \le x \le x_{i+1}: \phi_i(x)=\frac{x_{i+1}-x}{x_{i+1}-x_i}$$

$$x_{i+1} \le x \qquad : \phi_i(x)=0$$

---

[10] Also called "shape functions."

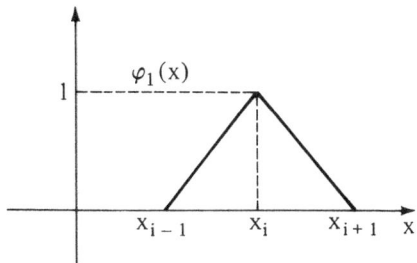

Fig. 1 (Note that $\phi_i(x_j)=\delta_{ij}$)

Thus $V_h$ is of dimension $N$: we can expand $u_h$ as:

$$u_h(x)=\sum_{i=1}^{N} U_i\phi_i(x)$$

with $u_i=$ value of $u_h$ at node $x_i=u_h(x_i)$

Finally the discrete equations (11) take the form:

---

Find $u_1,\ldots,u_N$ such that

$$\sum_{i=1}^{N} u_i a(\phi_i,\phi_j)=(f,\phi_j), \quad \text{for all } j=1,\ldots,N \qquad (12)$$

---

(that is, a system of $N$ linear equations for $N$ unknowns).

It is fruitful exercise, once in life, to compute the coefficients in (12) (using the quadrature formula:

$$\int_{x_i}^{x_{i+1}} \mu uv\, dx \simeq \mu_{i+1/2}\int_{x_i}^{x_{i+1}} uv\, dx )$$

One easily gets: $j=1,\ldots,N$:

$$a_{j,j-1}=\frac{-\kappa_{j-1/2}}{x_j-x_{j-1}}+\mu_{j-1/2}\frac{x_j-x_{j-1}}{6} \qquad (13a)$$

$j=1,\ldots,N-1$:

$$a_{j,j}=\frac{\kappa_{j-1/2}}{x_j-x_{j-1}}+\frac{\kappa_{j+1/2}}{x_{j+1}-x_j}+\mu_{j+1/2}\frac{x_{j+1}-x_j}{3}+\mu_{j-1/2}\frac{x_j-x_{j-1}}{3} \qquad (13b)$$

$$a_{j,j+1}=\frac{-\kappa_{j+1/2}}{x_{j+1}-x_j}+\mu_{j+1/2}\frac{x_{j+1}-x_j}{6} \qquad (13c)$$

$j=N$

$$a_{N,N}=\frac{\kappa_{N-1/2}}{x_N-x_{N-1}}+\mu_{N-1/2}\frac{x_N-x_{N-1}}{3} \qquad (13d)$$

(all other terms vanish).

For a uniform mesh: $x_j - x_{j-1} \equiv h$ and constant coefficients, equations take a familiar form:

$$-\kappa \frac{u_{i-1} + u_{i+1} - 2u_i}{h^2} + \mu \left( \frac{u_{i-1}}{6} + \frac{2u_i}{3} + \frac{u_{i+1}}{6} \right) = f_i$$

$$\left( \text{with } f_i = h^{-1} \int_0^1 f\phi_i \, dx \right) \quad (14)$$

$$u_0 = 1$$

$$-\kappa \frac{u_{N-1} - u_N}{h^2} + \mu \left( \frac{u_{N-1}}{6} + \frac{u_N}{3} \right) = f_N$$

As often the finite element methods of low degree, when applied on a uniform mesh, coincide with a standard finite difference scheme.

We note that the matrix of the linear system (12) is symmetric and tridiagonal; therefore the system of equations can be solved by Gauss elimination in $O(N)$ operations. Note also that the matrix is positive definite.

We sum up the general *features* of finite element methods which have been outlined in this section and which are common to all finite element methods:

The method is based on a weak formulation;

Dirichlet[11] boundary conditions are strongly imposed;

The unknowns of the problem have an immediate physical meaning: $u_i = u_h(x_i)$; (these are the "*degrees of freedom*" of $u_h$);

The basis functions $\phi_i$ have a small support[12] ($\phi_i$ is non vanishing only on the two intervals near $x_i$);

As a result, the matrix of the system to be solved is *sparse*: most of its elements are zero (in this 1D example it was even tridiagonal).

What about non homogeneous boundary condition? Suppose first that instead of (2) we have:

$$u(0) = u_0$$

Then we can set:

$$\tilde{u}(x) = u(x) - u_0\psi(x), \quad \text{where } \psi \text{ is such that } \psi(0) = 1$$

and the weak formulation is changed to:

$$a(\tilde{u}, v) = (f, v) + a(u_0\psi, v), \qquad \forall v$$

In the discrete problem we can take:

$$\tilde{u}_h(x) = u_h(x) - u_0\phi_0(x) \in V_h$$

but the equations (13)–(14) are unchanged: instead of vanishing $u_0$ takes a prescribed value.

---

[11] Boundary conditions of order $\leq m$ for an elliptic equation of order $2m$.

[12] Contrary to spectral methods, we have a *grid* point approximation; the degrees of freedom have a *local* signification.

Suppose now that we have a Fourier boundary condition at $x=1$:

$$u(0)=0$$

$$\alpha u(1)+\beta\frac{du}{dx}(1)=q_1, \qquad (\alpha>0, \beta>0) \tag{15}$$

From (5) it is readily seen that the corresponding weak formulation is:

$$\int_0^1\left(\kappa\frac{du}{dx}\frac{dv}{dx}+\mu uv\right)dx+\frac{\kappa(1)\alpha}{\beta}u(1)v(1)=\int_0^1 fv\,dx+\frac{\kappa(1)}{\beta}q_1v(1),$$

$$\forall\,v\in V. \tag{16}$$

The finite element equations are constructed as before from the weak formulation.

It is a common feature of finite element methods that the most general *boundary conditions* can be easily handled in a semi automatic fashion.

**Remarks.**

(i) The "energy" $\frac{1}{2}a(v,v)$ is augmented by a boundary term: $\kappa(1)\alpha/2\beta v(1)^2$

(ii) Neumann (or Fourier) boundary conditions are not strongly satisfied. For example, take $\kappa=1$, $\mu=0$, $f\equiv1$, and the *Neumann* condition at $x=1$

$$\frac{du}{dx}(1)=0.$$

Then the discrete solution is:

$$u_i=x_i-\frac{x_i^2}{2}, \qquad u_h=\sum_i u_i\phi_i$$

(The values of $u_h$ at the nodal points $x_i$ are *precisely* the values of the *exact* solution at $x_i$).[13] However the Neumann condition is only approximately satisfied, in the limit $h\to0$:

$$\frac{du_i}{dx}(1)=h/2$$

## 1.2. A 2-Dimensional Problem

We consider the 2-dimensional analogue of equation (1)

$$-(\kappa_{ij}u_{,i})_{,j}=f \quad \text{in } \Omega \tag{17a}$$

$$u=0 \quad \text{on } \Gamma_0, \quad \text{(Dirichlet boundary condition)} \tag{17b}$$

$$\kappa_{ij}u_{,i}n_j=0 \quad \text{on } \Gamma_1, \quad \text{(Neumann boundary condition)} \tag{17c}$$

---

[13] This is related to "superconvergence" phenomena: see Douglas (1972), Douglas and Dupont (1972).

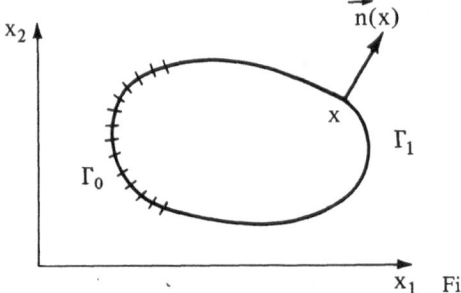

Fig. 2

The following *notations* are used:

$\Omega$ = domain of integration (Figure 2)

$\Gamma = \Gamma_0 \cup \Gamma_1$ = boundary of $\Omega$

$\mathbf{n} = \mathbf{n}(x)$ = *unit* vector, *normal* to $\Gamma$ at point $x$, pointing outwards, with components $n_1, n_2$

$$u_{,i} = \frac{\partial u}{\partial x_i}$$

$\kappa_{ij}(x)$ = diffusivity tensor, satisfying the coercivity condition ($\kappa_0 > 0$)

$$\kappa_{ij}(x)\xi_i\xi_j \ge \kappa_0(\xi_1^2 + \xi_2^2), \quad \text{for all} \quad x \text{ in } \Omega, \tag{18}$$
$$\text{and all vectors } (\xi_1, \xi_2)$$

We shall also assume that the $\kappa_{ij}$ are bounded.

(We use throughout the classical condition on repeated indices; for instance $\kappa_{ij}u_{,i}$ is the same as:

$$\sum_{i=1}^{2} \kappa_{ij}u_{,i} = \kappa_{1j}u_{,1} + \kappa_{2j}u_{,2})$$

EXAMPLE. $\kappa_{11} = \kappa_{22} = 1$, $\quad \kappa_{12} = \kappa_{21} = 0$

This is Poisson's equation with mixed boundary conditions:

$$-\Delta u = f \quad \text{in } \Omega$$

$$u = 0 \quad \text{on } \Gamma_0$$

$$\frac{\partial u}{\partial n} = 0 \quad \text{on } \Gamma_2$$

The first step in the application to (17) of a finite element method, is to write the weak formulation of (17). The procedure is very much the same as in 1 dimension: let $v(x)$ be any "smooth enough" function satisfying the homogeneous Dirichlet boundary condition (17b):

$$v \equiv 0 \quad \text{on } \Gamma_0. \tag{19}$$

Multiplying (17a) by $v$, integrate the result over $\Omega$ and transform it by Green's formula.[14] The result is:

$$\int_\Omega \kappa_{ij} u_{,i} v_{,j} \, dx = \int_\Omega f v \, dx. \tag{20}$$

(The boundary integrals vanish, with (17c) and (19).)

We require (20) to be true over all $v$ such that the integrals are meaningful:

$V = $ set of $v$ such that

a) $\int_\Omega |v|^2 \, dx < +\infty$

b) $\int_\Omega |v_{,i}|^2 \, dx < +\infty$         (21)

c) $v|_{\Gamma_0} = 0.$[15]

Note again the difference in the treatment of Dirichlet and Neumann boundary conditions: the former are strongly imposed by $u \in V$, while the latter is known to be satisfied only by solution $u$.

Finally the weak formulation is:

$$\boxed{\begin{aligned} & u \in V \\ & a(u,v) = (f,v), \quad \forall\, v \in V \end{aligned}} \tag{22}$$

with

$$\boxed{\begin{aligned} & a(u,v) = \int_\Omega \kappa_{ij} u_{,i} v_{,j} \, dx \\[2mm] & (f,v) = \int_\Omega f v \, dx \end{aligned}} \tag{23}$$

---

14 This is the 2-dimensional integration by parts:

$$\int_\Omega (\phi_{,i} \psi + \phi \psi_{,i}) \, dx = \int_\Gamma \phi \psi n_i \, d\Gamma.$$

15 If $\Gamma_0$ is reduced to a point (or has zero measure) (21-c) is meaningless, and we have in fact a pure Neumann problem; $u$ is then defined up to an additive constant.

**Remarks.**

(i) The above problem is, in general, not an optimisation problem, unless the tensor $\kappa_{ij}$ is symmetric: $\kappa_{ij} = \kappa_{ji}$; then $a(u, v) = a(v, u)$, and (22) is equivalent to find in $V$ the minimum of the functional $\frac{1}{2} a(v, v) - (f, v)$

(ii) The role of the coercivity condition (18) is to ensure existence and uniqueness of a solution (thanks to the Lax Milgram theorem, see e.g. Lions (1962), Agmon (1965)). It implies that:

$$\int_\Omega \kappa_{ij} v_{,j} v_{,j} \, dx \geq \kappa_0 \int_\Omega v_{,i} v_{,i} \, dx$$

In the symmetric case, $\kappa_{ij} = \kappa_{ji}$, $a(v, v)^{1/2}$ can be used as a norm on $V$ equivalent to the usual Sobolev norm.


## 1.3. The Finite Element Equations

From (22) the principle of finite element methods is straight forward: develop $u_h$ [16] (the discrete solution) as:

$$u_h(x) = \sum_{p=1}^{N} u_p \phi_p(x), \tag{24}$$

where the $\phi_p$ are the *basis* functions and $u_i$ the degrees of freedom of $u_h$.

Then the discrete equations are:

$$\sum_{p=1}^{N} u_p a(\phi_p, \phi_q) = (f, \phi_q), \qquad \forall q \ (q = 1, \dots, N). \tag{25}$$

We note $V_h$ the vector space generated by the functions $\phi_p$ ($p = 1, \dots, N$).

The above expression is strictly speaking meaningful if $\phi_p \in V$ ($V_h \in V$); however we will give later an example where this is not verified.

Thus the finite element method is entirely defined by the choice of the interpolating space $V_h$, that is to say, by the choice of the basis functions $\phi_p$.

The domain $\Omega$ is divided into small pieces, or "*elements*": intervals in 1-D problems; triangles or quadrangles in 2-D; tetrahedra or "bricks" in 3-D.[17] The basis functions are then defined element by element: we give below several examples where they are *polynomials* of moderate degree $k$ ($k = 1$ or 2).

---

[16] usually "$h$" is of the order of the meshsize.

[17] In 3-D, it seems much simpler to cover the domain with bricks than tetrahedra.

We shall use the notation:

$P_k$ = set of polynomials of degree $\leq k$

Thus:

$$P_0 = \{1\}$$
$$P_1 = \{1, x_1, x_2\}$$
$$P_2 = \{1, x_1, x_2, x_1^2, x_1 x_2, x_2^2\}, \text{ etc.}$$

When quadrangles are employed, the following spaces of polynomials are used:

$Q_k$ = set of polynomials of degree $\leq k$, with respect to *each* variable

$$Q_0 = \{1\}$$
$$Q_1 = \{1, x_1, x_2, x_1 x_2\}$$
$$Q_2 = \{1, x_1, x_2, x_1^2, x_1 x_2, x_2^2, x_1^2 x_2, x_2^2 x_1\}$$

We shall consider finite elements of *Lagrange* type, where the degrees of freedom are the values of the unknown function (on the contrary, in *Hermite* finite elements, the degrees of freedom also include values of derivatives).[18] [19]

In other words, $\phi_p$ is defined by its values at some specially chosen *nodal points*: for instance, a polynomial of degree 2 is entirely defined by its values at 6 different points. The choice of the nodal points is guided by *continuity* requirements: indeed we want $\phi_p$ (and $u_h) \in V$ (except in the case of non conforming elements, §1.) a sufficient condition is that:

(a) $\phi_p$ is piecewise polynomial
(b) $\phi_p$ is continuous at the inter-element faces                                (26)
(c) $\phi_p = 0$ on $\Gamma_0$

The continuity requirement is quite simple to check in the examples given in the following sections.

As to (26-c) it is also easily verified when $\Omega$ is a polygon: just cancel the degrees of freedom corresponding to those nodes which lie on $\Gamma_0$. Of course this is not always the case in engineering problems: if we use straight triangles, a "skin" remains between the exact domain and its approximation as shown on Figure 3. Thus an *error* is committed, which deteriorates the accuracy in finite elements of degree $\geq 2$.

This is the reason for the introduction of "*curved*" or "*isoparametric*" finite element methods (§1-4-2), where the boundary is better approximated.

---

[18] For instance we shall not consider in this course the Argyris' triangle, with polynomials of degree $\leq 5$: see Olson (1978) for application to Navier Stokes equations.

[19] In section 1.5 we shall introduce the *mixed formulation*, where the derivatives $u_{,i}$ are taken as independent variables, to be interpolated by another set of basis functions. In Hermite finite elements, the functions to be interpolated is still $u$.

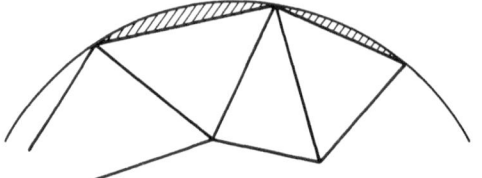

Fig. 3

In section 1-4-2 we shall consider the computation of integrals involved in (23): numerical integration is required when $\kappa_{ij}$ is not constant, or when curved elements are used; in any case it is recommended for elements of degree $\geq 2$. The question is: which quadrature formula should be used so that: i) a minimum of integration points are involved; ii) the accuracy is preserved.

*Error estimates.* We have the following general "theorem": provided the elements do not degenerate (in a sense to be precised), when polynomials of degree $k$ are used:

$$\left(\int_\Omega |\mathrm{grad}(u-u_h)|^2\,dx\right)^{1/2} = O(h^k)$$

$$\left(\int_\Omega |u-u_h|^2\,dx\right)^{1/2} = O(h^{k+1})$$

(27)

(Zlamal (1968), Srang and Fix (1973), Ciarlet and Raviart (1972), Ciarlet (1975), Raviart (1975)).

This "optimal" order of convergence is obtained under several conditions:
(i) The first is related to the possible degeneracy of elements: in the above papers, all the angles in the triangles or quadrangles are required to be bounded away from 0 *and* from $\Pi$. P. Jamet (1975, 1976) and Synge (1957) relaxed this condition and proved that the angles only need to be kept away from $\Pi$; thus triangles may degenerate into flat triangles with no obtuse angle (Figures 4 and 4b), quadrangles may degenerate into triangles (Figure 4c).

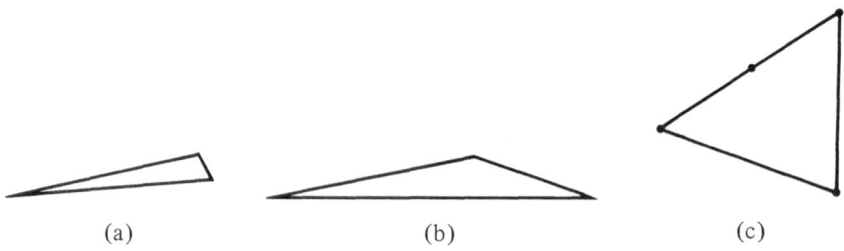

(a)                          (b)                          (c)

Fig. 4 (a) admissible degeneracy of triangles; (b) nonadmissible degeneracy of triangles; (c) admissible degeneracy of quadrangles

**Remark.** The price to be paid when using almost degenerate elements (Figure 4a) is an increased condition number of the matrix to be inverted. As an illustration, we give[20] a numerical application due to J. F. Bourgat, which conforts Jamet's analysis (Figs. 4b–4c and 9c–9e).

(ii) Next, the accuracy depends on the *smoothness* of the solution $u$: in particular the accuracy deteriorates when $u$ contains singularities (e.g. near corners in $\Omega$, or in ase of discontinuous $\kappa_{ij}$ or $f$). In such cases, the accuracy may be restored:

Either by a refined mesh near the singularity point (Cf. Becker and Carey (1978). Note that these authors relax the classical condition that two neighbouring elements should intersect either at a single point, or a whole side; see also Lomax (1977)).

Or by the use of special basis functions having the same order of singularity as $u$ (when this can be determined) at the expense of increased complexity (especially in the numerical integration). For more details, see Strang and Fix (1972, chapter 8).

(iii) We must use a proper quadrature formula and a proper representation of the boundary (if necessary, curved elements): we detail these points for the $P2$-triangle, §1-4-2.

(iv) The $O(h^k)$ accuracy (in the energy norm) may possibly not hold in hyperbolic problems of fluids mechanics, as we shall see in the following chapters.

## 1.4. Standard Examples of Finite Element Methods

This list of examples is not designed to be exhaustive of all possible finite element methods: see Zienckiewicz (1971) or Ciarlet (1975) for more examples: we only want to give illustrative examples that will appear to be useful in fluid mechanics applications (in chapters 4 and 5).

### 1.4.1. Example 1: The $P1$-Triangle (Courant's Triangle)

This is historically the first of finite element methods, since the pioneering paper by R. Courant (1943).

The discrete solution $u_h$ is piecewise polynomial of degree $\leq 1$: on each triangle, it is determined by its values at the 3 vertices. The continuity requirement on $u_h$ is checked by a straight forward argument: consider two adjacent triangles $T$ and $T'$ with interface $(ab)$ (Figure 5). The restriction of

---

[20] in appendix 2.

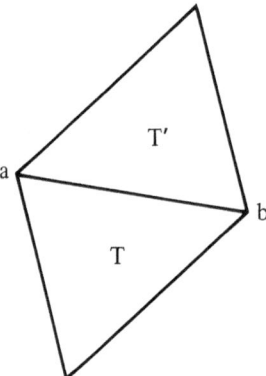

Fig. 5 Two adjacent elements

$u_h$ to $(ab)$ is a polynomial of degree $\leq 1$ in one variable; it depends on its values at the 2 end points $a$ and $b$; $u_h$ is continuous at these 2 points since $u_h(a)$ and $u_h(b)$ are nodal values. Therefore, on $(ab)$, $u_h$ takes the same value on each side, computed either from $T$ or from $T'$.

Each *basis function* $\phi_i$ is associated with a *vertex* node

$$a_i: \phi_i = 1 \text{ at } a_i,$$
$$\phi_i = 0 \text{ at all other nodes.}$$

Of course $\phi_i$ takes non zero values only on the triangles neighbour to vertex $a_i$ on a triangle $T$, whose vertices are numbered for convenience: $a_1, a_2, a_3$, the basis function simply coincide with the *barycentric* coordinates, or "*area coordinates*":

$$\phi_1(x) = \lambda_1(x) = \frac{\text{area of triangle } (xa_2a_3)}{\text{area of triangle } (a_1a_2a_3)}$$

The gradient of $\phi_1$ has a simple expression: (on triangle $T = a_1a_2a_3$, see Figure 6)

$$\text{grad } \phi_2 = -\frac{\mathbf{n}_1|a_2a_3|}{2\,\text{area}(T)}$$

$$= -\frac{1}{2\,\text{area}(T)}(a_3^2 - a_2^2) + \frac{1}{2\,\text{area}(T)}(a_3^1 - a_2^1) \qquad (28)$$

Let us take now a global view at the whole domain: for each node $p$ [21] there is an associated basis function $\phi_p$; the discrete equations are written from (25):

$$\sum_{p=1}^{N} u_p \times \sum_{T} \int_{T} \kappa_{ij}\phi_{p,i}\phi_{q,j}\,dx = \sum_{T}\int_{T} f\phi_q\,dx, \quad \text{for all } q = 1,\ldots,N \qquad (29)$$

---

[21] We number $N+1, N+2,\ldots$, the nodes on $\Gamma_0$.

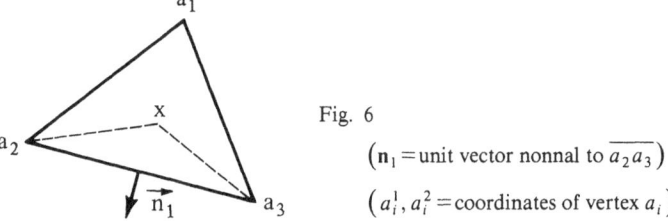

Fig. 6

$\left( \mathbf{n}_1 = \text{unit vector nonnal to } \overline{a_2 a_3} \right)$

$\left( a_i^1, a_i^2 = \text{coordinates of vertex } a_i \right)$

The matrix of this linear system of equations is of order $N=$ number of vertex nodes which do *not* lie on $\Gamma_0$ (where Dirichlet boundary conditions are applied).

The linear system is assembled from the contributions in the mesh, which are examined in turn.

An entry in the matrix: $a(\phi_p, \phi_q)$ (for any nodes $p, q$) is non zero only if $p$ and $q$ belong to a common triangle. Thus the matrix is sparse, but its structure is in general more complicated than matrices of finite difference methods, although it may be simplified when a regular mesh is used (see Figure 7: with this regular mesh, and if $\kappa_{11} = \kappa_{22} = 1$, $\kappa_{12} = \kappa_{21} = 0$, the classical 5-point finite difference formula for the Laplacian is retrieved).

Consider now the Poisson's equation $(-\Delta u = f)$: on a general mesh, it is easy to check from (28) that: provided the angles in the mesh are all *acute*, then:

$$a(\phi_p, \phi_p) > 0, \qquad a(\phi_p, \phi_q) \le 0, \qquad q \neq p \tag{30}$$

$$\sum_q a(\phi_p, \phi_q) \ge 0$$

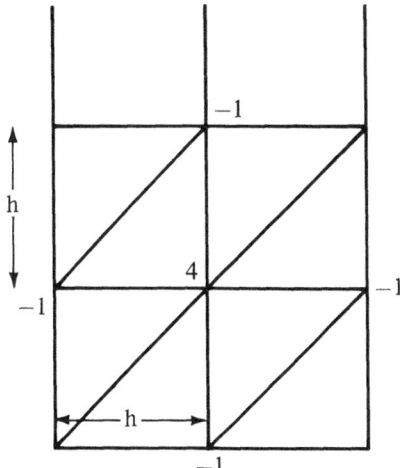

Fig. 7

Then the matrix is an $M$-matrix (Varga (1962)), that is, its inverse has positive elements; thus,

$$f \geq 0 \Rightarrow u_h \geq 0 \qquad (31)$$

This discrete maximum principle is lost for higher order finite elements.
    Cf. Ciarlet and Raviart (1973b).


### 1.4.2. Example 2: The $P2$-Triangle

The discrete solution $u_h$ is piecewise *quadratic* and *continuous* all over $\Omega$. On each triangle $T = (a_1 a_2 a_3)$ (Figure 8: the mid points are $a_4, a_5, a_6$), there are 6 basis functions whose expression is easily found with the help of barycentric coordinates:

One basis function per vertex: $\phi_{a_1} = \lambda_1 (2\lambda_2 - 1)$
One basis function per midside node: $\phi_{a_4} = 4\lambda_1\lambda_2$

**Remarks.**
    $\phi_{a_i}(a_j) = \delta_{ij}, \qquad i, j = 1, \ldots, 6.$
The discrete solution is continuous across an inter triangle boundary (by the same argument as in 1.4.1)

    If the domain of integration $\Omega$ is a polygon, the error in the energy norm is $O(h^2)$, after (27). If this is not true ($\Omega$ has curved boundaries), errors are committed when $\Omega$ is approximated by a polygon; then it can be proved (and confirmed by experience) that the error in energy norm cannot be better than $O(h^{3/2})$ (see e.g. Raviart (1975)). This is rather frustrating: why should we pay an increased complexity for an improvement of only $O(\sqrt{h})$? Hence appears the interest of a better approximation by *curved elements* which we now describe.
    Consider Figure 9a, where a triangle $T = (a_1 a_2 a_3)$ is close to a *curved* boundary: it would be nice to be able to interpolate the solution at point $a_5$, situated on the boundary, instead of the true mid point $\tilde{a}_5$; but how can we define a solution satisfying the continuity requirements? The trick is to transform $T$ into a straight triangle $\hat{T}$, and to define the basis functions on $\hat{T}$.

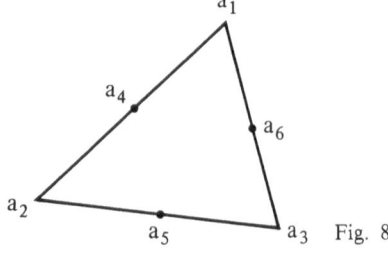

Fig. 8

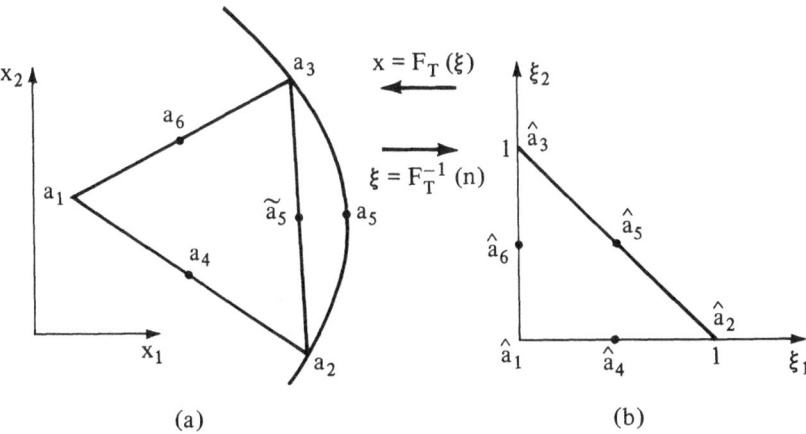

(a)                                                        (b)

Fig. 9 (a) a curved $P2$-triangle;   (b) the reference $P2$-triangle

It appears convenient to use the *same* $\hat{T}$ for all triangles $T$ in the mesh; the choice of $\hat{T}$ (the *reference* triangle) is just a matter of personal convenience for the programmer. A common choice is shown on Figure 9b. The transformation $F_T$, which is to map $\hat{T}$ onto $T$, should satisfy:

$$F_T(\hat{a}_l)=a_l, \qquad l=1,\ldots,6 \tag{32}$$

That is to say, the image of a node in the reference element is a node in the triangle $T$; note that, for $l=4,5,6$, $a_l$ does not need to be the true mid point $\tilde{a}_l$ of the straight segment event if $a_l$ is not on the boundary (e.g. $a_6$ or $a_4$ on Figure 9a); however the distance between $a_i$ and $\tilde{a}_i$ should be by an order of magnitude smaller than the meshsize:

$$\text{if } a_1a_2=O(h), |a_5\tilde{a}_5|=O(h^2). \tag{33}$$

Now (32) states, for each cartesian component of $F_T$, 6 equations: therefore we assume for $F_T(\xi_1,\xi_2)$ is quadratic, and (32) entirely determines $F_T$.

It is quite convenient to define $F_T$ with the basis functions of the reference element:

$$\hat{\phi}_l(\hat{a}_m)=\begin{cases} 1, & \text{if } l=m \\ 0, & \text{if } l\neq m \end{cases} \qquad l,m=1,\ldots,6$$

Then:

$$F_T(\xi)=\sum_{l=1}^{6} a_l\hat{\phi}_l(\xi)$$

It is an elementary exercise to obtain the $\hat{\phi}_l$; for instance: $\hat{\phi}_1=(1-(\xi_1+\xi_2))(1-2(\xi_1+\xi_2))$, etc.

Now the basis functions $\phi_l$ on $T=F_T(\hat{T})$ are defined through $F_T$:

$$\phi_l(x)=\hat{\phi}_l(\xi), \qquad x=F_T(\xi), \qquad l=1,\ldots,6$$

The integrals on $T$ are transformed into integrals on $\hat{T}$:

$$\int_T \kappa_{ij}\phi_{l,i}\phi_{m,j}\,dx = \int_{\hat{T}} J_T \kappa_{ij}(F_T(\xi))\hat{\phi}_{l,r}\hat{\phi}_{m,s}\frac{\partial \xi_r}{\partial x_i}\frac{\partial \xi_s}{\partial x_j}\,d\xi \tag{34}$$

where $J_T = \det \dfrac{\partial F_T}{\partial \xi} = $ jacobian of $F_T$. (Note that $J_T(\xi)$ is a rational fraction).

**Remarks.**

(i) The same polynomial set is used to represent the geometry of the triangle and the solution: hence comes the qualification of "isoparametric" which is often applied to curved elements;

(ii) Under such reasonable hypotheses as (32), the jacobian of $F_T$ does not vanish and $F_T$ is an invertible mapping from $T$ onto $\hat{T}$;

(iii) Continuity requirements are checked by the standard argument developed in 1.4.1 for the $P$1-triangle (first, by the same argument, it can be checked that the curved elements cover the interior of $\Omega$ without overlapping). *Note*: a polynômial of degree 2 in 1 variable is determined by its values at 3 different points!

(iv) Clearly the integrals involved in (33) should be computed through *numerical integration*:

$$\int_{\hat{T}} g(\xi)\,d\xi \text{ is approximated by } \sum_{i=1}^{I} \omega_i g(\xi^i)$$

$$\left(\text{where the } \xi^i \text{ belong to } \hat{T}\right). \tag{35}$$

We see the capital interest in the choice of a common reference element: the numerical integration formula is fixed and does not need to depend on $T$. Which integration formula should be chosen? First of all the matrix and the right hand side may be computed through different integration formulae; then it is expected than an extra cost will have to be paid when curved elements are used instead of the straight ones.

A balance must be kept between accuracy and computer time: a very accurate integration formula will be costly; on the other hand, (34) must be accurate enough not to deteriorate the $O(h^k)$ error in the energy norm. From the analysis of Ciarlet and Raviart (1972) it results that, when $P$2-curved triangles are used, the following sufficient conditions should be satisfied:

(i) $\omega_i > 0$, $i = 1, \dots, I$

(ii) There are enough integration points in (34) so that the approximation of $a(u, v)$ is non degenerate (it should define a coercive bilinear form on $V_h$): at least 3 integration points should be used;

(iii) For the computation of (34), the integration formula (35) should be exact for polynomials of degree $\leq 3$ ($\leq 2$ if the triangle is straight);

(iv) For the computation of the right hand sides $\int_T f\phi_j\,dx$, the integration formula should be exact for polynomials of degree $\leq 4$ ($\leq 2$ for a straight triangle).

**Remark.** This analysis holds if the data $f$ and $\kappa_{ij}$ are smooth; for instance $\kappa_{ij}$ should have, on each triangle, bounded derivatives; if this is not true the true solution $u$ will show singularities, and the optimal order of accuracy will not be obtained any way.

All these remarks on numerical integration stay valid, in their principle, for the subsequent examples.

Numerical illustration (J. F. Bourgat, appendix 2).

### 1.4.3. Example 3: The $Q1$-Quadrangle

Although this is the simplest quadrangular element, the passage to a reference element is required where the transformation has non constant jacobian.[22]
The basis functions on the reference element $\hat{K}$ (Figure 10b) are:

$$\hat{\phi}_1 = (1-\xi_1)(1-\xi_2)$$
$$\hat{\phi}_2 = \xi_1(1-\xi_2)$$
$$\hat{\phi}_3 = \xi_1\xi_2$$
$$\hat{\phi}_4 = \xi_1(1-\xi_2)$$

so that $\hat{\phi}_l(\hat{a}_m)=\delta_{lm}$.

For a quadrangle $K=a_1a_2a_3a_4$ (Figure 10a—$K$ should be *convex* in order to ensure that $F_K$ is non singular) we set

$$F_K(\xi)= \sum_{l=1}^{4} \hat{\phi}_l(\xi)a_l$$

Next we define the basis functions on $K$:

$$\phi_l(x)=\hat{\phi}_l(\xi), \qquad x\in K, \quad \xi\in\hat{K}, \quad x=F_K(\xi).$$

We do not detail the development of the method which is quite similar to the other examples.

Order of accuracy:

$$\left( \int_\Omega |grad(u-u_h)|^2\, dx \right)^{1/2} = O(h).$$

Integration formula:

Contains at least three points;
Exact for all polynômials in $Q_1$.

---

[22] Unless one is content with working on simple domains: the image of the square $\hat{K}$ under a linear transformation is a parallelogram, it cannot be a general quadrangle of arbitrary shape.

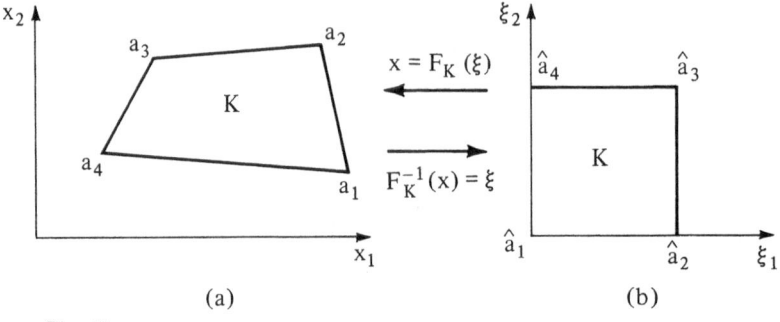

Fig. 10

### 1.4.4. Example 4: The $Q2$-Quadrangle

Basis functions on $\hat{K}$:

$$\hat{\phi}_1 = (1-\xi_1)(1-\xi_2)(1-2\xi_1)(1-2\xi_2)$$

$$\hat{\phi}_2 = \xi_1(1-\xi_2)(1-2\xi_1)(1-2\xi_2)$$

$$\hat{\phi}_3 = \xi_1\xi_2(1-2\xi_1)(1-2\xi_2)$$

$$\hat{\phi}_4 = (1-\xi_1)\xi_2(1-2\xi_1)(1-2\xi_2)$$

$$\hat{\phi}_5 = 4\xi_1(1-\xi_2)(1-\xi_1)(1-2\xi_2)$$

$$\hat{\phi}_6 = 4\xi_1\xi_2(1-2\xi_1)(1-\xi_2)$$

$$\hat{\phi}_7 = 4\xi_1\xi_2(1-\xi_1)(1-2\xi_2)$$

$$\hat{\phi}_8 = 4(1-\xi_1)\xi_2(1-2\xi_1)(1-\xi_2)$$

$$\hat{\phi}_9 = 16\xi_1\xi_2(1-\xi_1)(1-\xi_2)$$

$$\xi \in K \rightarrow x \in \hat{K}: x = F_K(\xi) = \sum_{l=1}^{9} a_l\hat{\phi}_l(\xi)$$

$$\phi_l(x) = \hat{\phi}_l(\xi), \qquad x = F_K(\xi), \qquad l = 1,\dots,9$$

Order of accuracy:

$$\left( \int_\Omega |\mathrm{grad}(u-u_h)|^2\, dx \right)^{1/2} = O(h^2)$$

Integration formula:

A polynomial of $P_3 \cap Q_2$ is determined by its values at the integration points[23];

The integration formula integrates exactly polynomials in $Q5$ (in $Q3$ if the element is "straight", i.e. if $F_K \in Q_1 \times Q_1$).

---

[23] In the terminology of Raviart (1975), the integration points form a $P_3 \cap Q_2$ unisolvent set.

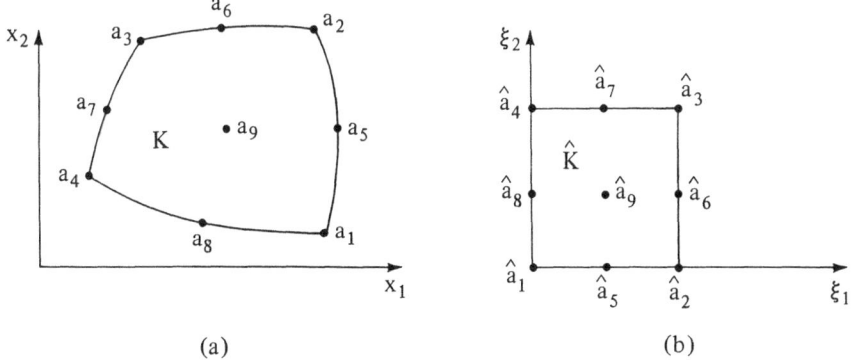

Fig. 11 (a) a curved $Q2$-quadrangle; (b) the reference $Q2$-quadrangle

**Remarks.** It is quite legitimate to mix in the same mesh $P_2$ (curved) triangles and $Q_2$ (curved) quadrangles, or $P_1$ triangles and $Q_1$ quadrangles, at the expense of minor increase in programming complexity.

### 1.4.5. A Variational Crime[24]: The $P1$ Nonconforming Element

In all of the previous examples, the approximate solution was a continuous function all over the domain, and:

$$u_h \in V_h \subset V$$

Now, with the present example we allow this continuity requirement to be relaxed: the elements are (straight) triangles, $u_h$ is piecewise polynomial of degree $\leq 1$, and only required to be *continuous at the mid side nodes*; in other words we chose to determine, on each triangle, $u_h$ by its values at the mid side nodes.

$$u_h(x) = \sum_{p=1}^{N} u_p \phi_p(x), \qquad u_p = u_h(p)$$

$\phi_p$ is the basis function associated with mid side node $p$:

$\phi_p$ is piecewise linear

$$\phi_p(p) = 1$$

$$\phi_p(q) = 0 \ (q \neq p)$$

The support of $\phi_p$ is shown on Figure 12: this is made of the 2 triangles adjacent to node $p$. Strictly speaking $a(\phi_p, \phi_q)$ is not defined: the derivatives of $\phi_p, \phi_q, u_h$, contain Dirac distributions; let us *ignore* them and write the

---

[24] After G. Strang.

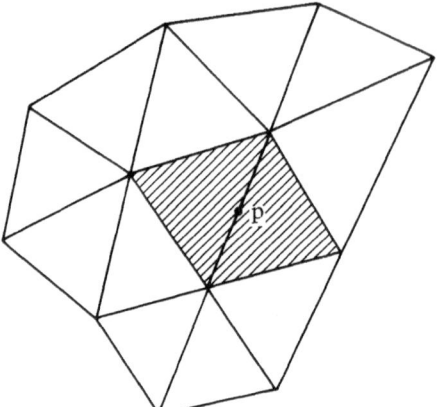

Fig. 12 The support of $\phi_p$

following discrete equations, analogue to (20);

$$\sum_{p=1}^{N} u_p \sum_T \kappa_{ij} \phi_{p,i} \phi_{q,j} \, dx = \sum_T \int_T f \phi_q \, dx, \quad \text{for all } q=1,\dots,N$$

A justification is required for this process which may appear somewhat arbitrary: in fact, nonconforming finite element methods sometimes work (the present one does!), sometimes they don't! A necessary criterion for convergence, the "*patch test*" is due to the intuition of B. M. Irons (1972); it was mathematically analysed by G. Strang (Strang and Fix (1972)). The patch test is enounced as follows:

> Suppose that the data $f, \kappa_{ij}$, are such that the exact solution is a polynomial $u$ [25] of degree 1 [26], then the approximate solution $u_h$ should coincide exactly with $u$.

This test is very simple to check; it proved to be successful in structural mechanics applications.

We give in the appendix a sketch of the proof of convergence (for a simplified equation) and of the error estimate in the energy norm:

$$\left( \sum_T \int_T |(u-u_h)_{,i}|^2 \, dx \right)^{1/2} = O(h)$$

Finally we note some further properties of the nonconforming $P1$-triangle:

(i)  There are about 3 more unknowns than with the $P1$ conforming triangle, for the same order of accuracy. However we shall see that this

---

[25] Not piecewise polynomial, but polynomial on the whole domain.

[26] Of degree $m$ for an equation of order $2m$.

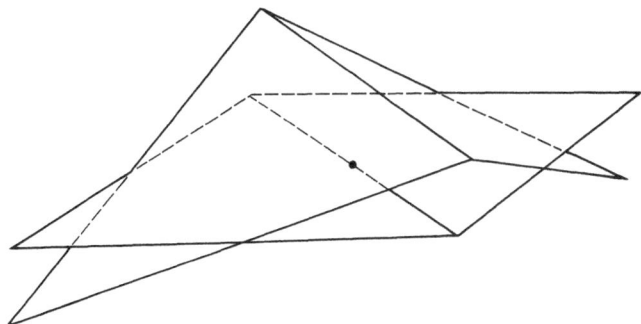

Fig. 13 Perspective
view of $\phi_p$ on the 2
adjacent triangles

element defines a cheap approximation of Stokes problem (chapter 4). For application to the approximation of shallow water equations, see Hua and Thomasset (1979);

(ii)  The properties (30) and (31) still hold for the approximation of Poisson's equation (discrete maximum principle);

(iii) If $p \neq q$, the basis function $\phi_p$ and $\phi_q$ are $L^2$-orthogonal:

$$\int_\Omega \phi_p \phi_q \, dx = 0, \quad \text{if } p \neq q$$

(this fails to be true for the $P1$ conforming triangle, unless the "mass-lumping" approximation is used);

(iv) The importance of such properties are enlightened by the examination of a parabolic equation, e.g. heat conduction equation in a homogeneous medium (with normalized coefficient):

$$\frac{\partial u}{\partial t} - \Delta u = f$$

$$u|_\Gamma = 0$$

$$u(0) = u_0$$

with $f \geq 0, u_0 \geq 0$

Discretizing (in space) with the $P1$ nonconforming triangle and in time with explicit schemes (like Euler), or implicit schemes (Crank Nicholson) it is easy to check that the approximate temperature *remains positive*.

## 1.5. Mixed Formulation and Mixed Finite Element Methods for Elliptic Equations

The word "mixed" refers to the fact that the *gradient* of the solution is introduced as a supplementary independent variable: thus one has to mix (with some caution) two types of approximations for $u$, and for grad $u$. Note that this approach is quite different of interpolating $u$ through values of $u$ and grad $u$ at nodal points (Hermite interpolation). The interest of such approach is two fold: first the accuracy for grad $u$ is increased (for instance Neumann boundary conditions are better satisfied); this is important if the variable of primary physical interest is grad $u$ rather than $u$ in itself.[27] Then, commonly used approximations of Stokes and Navier Stokes equations come into the framework of mixed finite element methods (velocity-pressure formulation, or stream function-vorticity), and new promising methods are being developed within this framework. The price to be paid is generally a considerable increase in the number of equations.

The aim of this chapter is to explain what mixed finite element methods really are and which necessary conditions should be satisfied for convergence. A detailed analysis of mixed anhybrid methods can be found in Thomas (1977); see also Thomas (1976), Raviart and Thomas (1977a, 1977b), Brezzi (1974), Brezzi and Raviart (1978), Johnson and Mercier (1978), and the references of these papers.

### 1.5.1. The One Dimensional Problem

In order to get a better understanding of the method, we consider again the one dimensional problem of chapter 1.

$$-\frac{d}{dx}\left(\kappa\frac{du}{dx}\right)=f \quad \text{in } )0,1($$
$$u(0)=0 \tag{36}$$
$$\frac{du}{dx}(1)=0$$

We can write (36) in the form:

a)  $\quad -\dfrac{dp}{dx}=f \quad \text{in } )0,1($

b)  $\quad p-\kappa\dfrac{du}{dx}=0 \quad \text{in } )0,1($   $\qquad$ (37)

c)  $\quad u(0)=0$

d)  $\quad p(1)=0$

A weak formulation corresponding to (37) is as follows:

a)  $\quad \displaystyle\int_0^1\left(\frac{dp}{dx}+f\right)v\,dx=0, \qquad \forall\,v\in L^2(\Omega)$

$\qquad$ (38)

b)  $\quad \displaystyle\int_0^1\left(\kappa^{-1}pq+u\frac{dq}{dx}\right)dx=0, \qquad \forall\,q\in X,$

---

[27] For instance: in irrational flows: $u=$ potential, grad $u=$ velocity.

where $X$ is the set of functions $q$ on $)0, 1($ such that:

$$\int_0^1 |q|^2 \, dx < +\infty$$

$$\int_0^1 \left|\frac{dq}{dx}\right|^2 dx < +\infty$$

$$q(1) = 0$$

(so that $p \in X$ implies the Neumann boundary condition (37d)).

**Remarks.**

(i) (38b) is nothing but a weak form of: $p - \kappa(du/dx) = 0$, with *boundary* condition $u(0) = 0$;

(ii) no derivatives of $u$ or $v$ appear in (38); therefore we can take a *discontinuous* approximation of $u$.

For brevity, we shall only describe the mixed method of lowest degree:

Piecewise constant approximation of $u$;
Piecewise linear approximation of $q$.

Thus we divide $)0, 1($ into $N$ intervals (the 1-D "elements").
We set, for this subdivision of $)0, 1($:

$M_h$ = Set of piecewise constant functions; a set of basis functions of $M_h$ is given by the functions $w_i$:

$$w_i = 1, \quad \text{on } )x_i, x_{i+1}($$
$$= 0, \quad \text{elsewhere}$$
$$\text{(Figure 14a)}$$

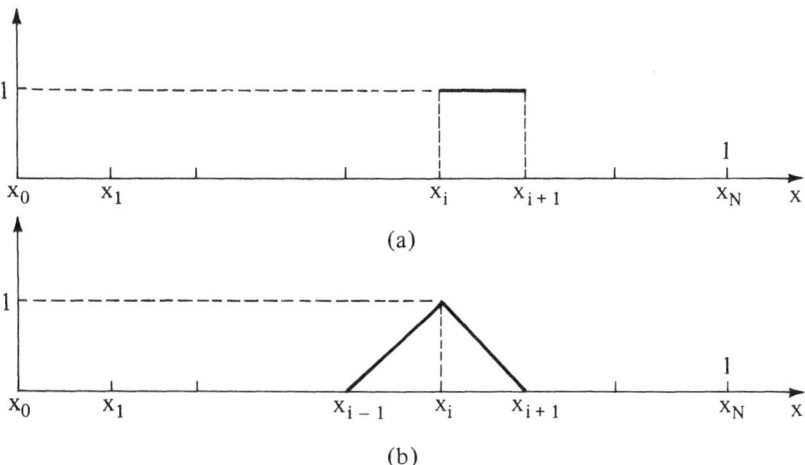

(a)

(b)

Fig. 14 (a) basis function of $M_h$;  (b) basis function of $X_h$

$X_h$ = Set of piecewise linear, continuous functions; vanishing at $x=1$ a set of basis functions of $X_h$ is given by the roof functions $\phi_i$ (figure 14b, $i=0,1,\ldots,N-1$)
(Note that $X_h \subset X$, $M_h \subset L^2(0,1)$).

Then, the discrete equations are readily obtained from (38):

$$\text{a)} \quad \int_0^1 \left( \frac{dp_h}{dx} + f \right) v\, dx = 0 \qquad \forall\, v \in M_h$$

$$\text{b)} \quad \int_0^1 \left( \kappa^{-1} p_h q + u_h \frac{dq}{dx} \right) dx = 0, \qquad \forall\, q \in X_h$$

$$(39)$$

(this last equation defines the *discrete derivative* of $u_h$) or, equivalently:

$$\text{a)} \quad \int_{x_i}^{x_{i+1}} \left( \frac{dp_h}{dx} + f \right) dx = 0, \quad i=0,1,\ldots,N-1$$

$$\text{b)} \quad \int_0^1 \left( \kappa^{-1} p_h q_i + u_h \frac{dq_i}{dx} \right) dx = 0, \quad i=0,\ldots,N-1$$

$$(40)$$

Let us assume for the sake of brevity that: $\kappa \equiv \kappa_0 > 0$ (this does not change the structure of the system of equations), and set:

$$\lambda_{i+1/2} = \frac{x_{i+1} - x_i}{3\kappa_0}$$

$$p_i = p_h(x_i), \qquad u_{i+1/2} = u_h \quad \text{on } )x_i, x_{i+1}($$

$$f_{i+1/2} = (x_{i+1} - x_i) \int_{x_i}^{x_{i+1}} f\, dx$$

Then the discrete equations can be explicitly written:

a) $\quad p_{i+1} - p_i = -f_{i+1/2}, \qquad i=0,1,\ldots,N-1$

b) $\quad p_N = 0$

c) $\quad \lambda_{1/2} p_0 + \tfrac{1}{2}\lambda_{1/2} p_1 - u_{1/2} = 0, \qquad (i=0)$

d) $\quad \tfrac{1}{2}\lambda_{i-1/2} p_{i-1} + (\lambda_{i-1/2} + \lambda_{i+1/2}) p_i + \tfrac{1}{2}\lambda_{i+1/2} p_{i+1}$
$\qquad + u_{i-1/2} - u_{i+1/2} = 0, \qquad (0 < i < N-1)$

e) $\quad \tfrac{1}{2}\lambda_{N-3/2} p_{N-2} + (\lambda_{N-3/2} + \lambda_{N-1/2}) p_{N-1} + u_{N-3/2} - u_{N-1/2} = 0,$
$\qquad (i = N-1)$

$$(41)$$

In matrix form we can write this as:

$$\left( \begin{array}{c|c} A & B^T \\ \hline B & 0 \end{array} \right) \left( \frac{p_0}{u_{1/2}} \right) = \left( \frac{0}{f_{1/2}} \right)$$

with

$$
A =
\begin{pmatrix}
\lambda_{1/2} & \tfrac{1}{2}\lambda_{1/2} & & & & \\
\tfrac{1}{2}\lambda_{1/2} & (\lambda_{1/2}+\lambda_{3/2}) & \tfrac{1}{2}\lambda_{3/2} & & 0 & \\
& & & & & \\
0 & & & & & \\
& & \tfrac{1}{2}\lambda_{i-1/2} & (\lambda_{i-1/2}+\lambda_{i+1/2}) & \tfrac{1}{2}\lambda_{i+1/2} & \\
& & & & & \tfrac{1}{2}\lambda_{N-3/2} \\
& & & \tfrac{1}{2}\lambda_{N-3/2} & (\lambda_{N-3/2}+\lambda_{N-1/2}) &
\end{pmatrix}
$$

$$(43)$$

$$
B =
\begin{bmatrix}
-1 & 1 & & & \\
& -1 & 1 & & 0 \\
& & & & 1 \\
0 & & & & -1
\end{bmatrix}
$$

(the fact that the dimensions of $A$ and $B$ are equal is particular to this 1-dimensional formulation).

Several important points should be emphasized:

(i) Some diagonal elements in the matrix of (42) are zero and the solver needs to take this fact into account[28];
(ii) There are twice as many unknowns than for the Lagrange approximation of $u$ by piecewise linear elements;
(iii) If we solve (41) for $p_i$ we find:

$$
p_i = p_h(x_i) = -\int_{x_i}^{1} f(x)\, dx,
$$

that is, the exact solution is obtained at the nodes.

---

[28] In this very simple problem, direct solution of (41) is straightforward, solving first in $p_i$ and substituting in (41c–e) to obtain the $u_{i+1/2}$.

On the other hand:

$$u_h(0)=u_{1/2}=\frac{x_1}{3\kappa_0}\left(p_0+\frac{1}{2}p_1\right)=O(h) \tag{44}$$

Thus the Neumann condition is strongly satisfied and the Dirichlet condition (on $u$) is only weakly satisfied.

For the convergence of this method (and of higher order methods), we refer e.g. to Thomas (1977) where it is proved that:

$$\left(\int_0^1\left(|p_h-p|^2+\left|\frac{d}{dx}(p_h-p)\right|^2+|u_h-u|^2\right)dx\right)^{1/2}=O(h)$$

The keystone of the proof lies in the verification of the following property, known as Brezzi-Babuška condition:

$$\forall\,v\in M_h,\ \underset{q\in X_h}{\text{Sup}}\ \frac{\int_0^1 v\frac{dq}{dx}dx}{\left(\int_0^1\left(q^2+\left|\frac{dq}{dx}\right|^2\right)dx\right)^{1/2}}\geq C\left(\int_0^1|v|^2\,dx\right)^{1/2} \tag{45}$$

(where $C$ is a constant, independent of $h$). It is important to see the meaning of this condition: if the left hand side inequality (45) vanishes for $v=u_h$:

$$\int_0^1 u_h\frac{dq}{dx}dx=0,\quad\text{for all }q\text{ in }X_h.$$

Hence from (39b), $\int_0^1\kappa^{-1}p_h^2\,dx=0$, and $p_h\equiv0$; and on the other hand, (45) implies $u_h\equiv0$. Thus, the implication of (45) is this:

> if the discrete derivative of $u_h$ is 0, then $u_h\equiv$ constant ($=0$ from boundary conditions).

In other words, (45) tests[29] the *consistency* of the approximation of derivatives: the interpolations of $u$ and $p=\kappa\,du/dx$ cannot be chosen independently of each other. In this particular case, (45) is easily verified (for any $v$ in $M_h$, set $q(x)=\int_0^1 v(\xi)\,d\xi$, then $q\in X_h$, and $dq/dx=v$). However, it is much more technical to check the same verification in 2- or 3-dimensional problems.

### 1.5.2. A Two Dimensional Problem

Consider the two dimensional boundary value problem of chapter 1:

a)  $-(\kappa_{ij}u_{,i})_{,j}=f,\quad$ in $\Omega\subset\mathbb{R}^2$

b)  $u=0,\qquad\qquad$ on $\Gamma_0$  $\tag{46}$

c)  $\kappa_{ij}u_{,i}n_j=0,\qquad$ on $\Gamma_1$

---

[29] from another point of view, (45) ensures that the system of equations has a unique solution.

Obviously this problem can be split as in §2-2 by the introduction of the cogradient $\mathbf{p}=(p_1, p_2)$:

a)   $p_{j,j}+f=0,$          in $\Omega$

b)   $p_j=\kappa_{ij}u_{,i},$          in $\Omega$

c)   $u=0,$          on $\Gamma_0$          (47)

d)   $p_jn_j\equiv\mathbf{p}\cdot\mathbf{n}=0,$   on $\Gamma_1$

We rewrite (47b) as:

$$\lambda_{ij}p_i=u_{,j},\tag{48}$$

i.e., $[\lambda_{ij}]=[\kappa_{ij}]^{-1}$.

By straightforward manipulations we get a weak formulation of (47) (48) which we write under the general form: $\mathbf{p}\in X, u\in M$:

$$\boxed{\begin{aligned}
\text{a)}\quad & a(\mathbf{p},\mathbf{q})+b(\mathbf{q},v)=0,\qquad \forall\,\mathbf{q}\in X\\[2mm]
\text{b)}\quad & b(\mathbf{p},v)=-\int_\Omega fv\,dx,\qquad \forall\,v\in M
\end{aligned}}\tag{49}$$

with

a)   $a(\mathbf{p},\mathbf{q})=\displaystyle\int_\Omega \lambda_{ij}p_iq_j\,dx$          (50)

b)   $b(\mathbf{q},v)=\displaystyle\int_\Omega q_{j,j}v\,dx=\int_\Omega \operatorname{div}\mathbf{q}\,v\,dx$

$$X=\left\{\mathbf{q}(x):\int_\Omega|\mathbf{q}|^2\,dx<+\infty,\ \int_\Omega|\operatorname{div}q|^2\,dx<+\infty,\ \mathbf{q}\cdot\mathbf{n}=0\ \text{on}\ \Gamma_1\right\}\tag{51}$$

$$M=L^2(\Omega)=\left\{v:\int_\Omega v^2\,dx<+\infty\right\}$$

From (51), it is convenient to approximate $u\in M$ by *discontinuous* functions, piecewise polynomial of some degree $k$; the chief difficulty lies in the construction of the proper approximation for $\mathbf{p}\in X$. We shall describe here the *triangle* approximation of lowest degree $(k=0)$ and refer to Thomas (1977) or Raviart (1979) for higher order approximations or approximations on quadrilaterals.

Thus for a given triangulation, the approximation space for $u$ is:

$M_h=\{v$ such that: on each triangle $T$, $v=$constant$\}$

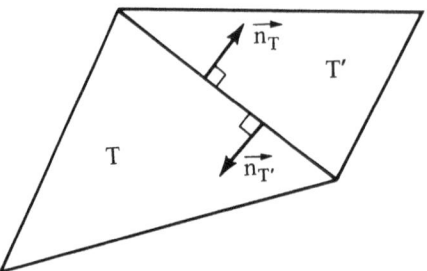

Fig. 15

The construction of the approximation of **p** lies on the following lemma:

> If **q** is smooth[30] on each triangle, a necessary and sufficient condition
> for $\int_\Omega |\mathrm{div}\, \mathbf{q}|^2\, dx < +\infty$ is the continuity of normal components of **q** at
> interfaces of triangles:
>
> $$\mathbf{q}|_T \cdot \mathbf{n}_T + \mathbf{q}|_{T'} \cdot \mathbf{n}_{T'} = 0 \quad \text{on } \partial T \cap \partial T'$$
>
> $(\mathbf{n}_T + \mathbf{n}_{T'} = 0, \text{ Figure 15}).$

$$(52)$$

Therefore quite naturally the degrees of freedom of $\mathbf{q} \in X_h$ should be the
normal components of **q** at some points on the triangle sides. It is convenient
to use the reference triangle $\hat{T}$ (Figure 16) to define the basis functions on $T$.

First we define the basis functions on $\hat{T}$, so that: $\hat{\phi}_i(\hat{m}_j) \cdot \hat{\mathbf{n}}^j = \delta_{ij}$:

$$\hat{\phi}_1(\xi) = \begin{vmatrix} \xi_1 \\ -1 + \xi_2 \end{vmatrix} \quad \hat{\phi}_2(\xi) = \begin{vmatrix} \xi_1\sqrt{2} \\ \xi_2\sqrt{2} \end{vmatrix} \quad \hat{\phi}_3(\xi) = \begin{vmatrix} -2 + \xi_2 = \hat{\phi}_{31} \\ \xi_2 = \hat{\phi}_{32} \end{vmatrix}$$

Then we define the basis functions on $T$:

$$\phi_i(x) = \frac{+1}{2 \cdot \mathrm{area}(T)} \frac{\partial F_T}{\partial \xi_r} \hat{\phi}_{ir}, \qquad i = 1, 2, 3 \tag{53}$$

which verify: $\phi_i(m_j) \cdot \mathbf{n}_T^j = \delta_{ij}$; in order to satisfy (52) we must change the
sign in (53) (since $n_T + n_{T'} = 0$ for 2 neighbor triangles). The approximation
space $X_h$ is defined to be generated by the basis functions $\phi_i$ associated to all
nodes in the mesh,[31] *but for* those nodes lying on $\Gamma_1$; then the discrete
equations are, from (49):

a)  $\quad a(\mathbf{p}_h, \mathbf{q}) + b(\mathbf{q}, u_h) = 0, \qquad \forall\, \mathbf{q} \in X_h \subset X$

b)  $\quad b(\mathbf{p}_h, v) = -\int_\Omega f \cdot v\, dx, \qquad \forall\, v \in M_h \subset L^2(\Omega).$

$$(54)$$

---

[30] $\mathbf{q}|_T \in H^1(T)$.

[31] For any element $\mathbf{q} \in X_h$, it can be checked that $\mathrm{div}\,\mathbf{q}$ is piecewise constant.

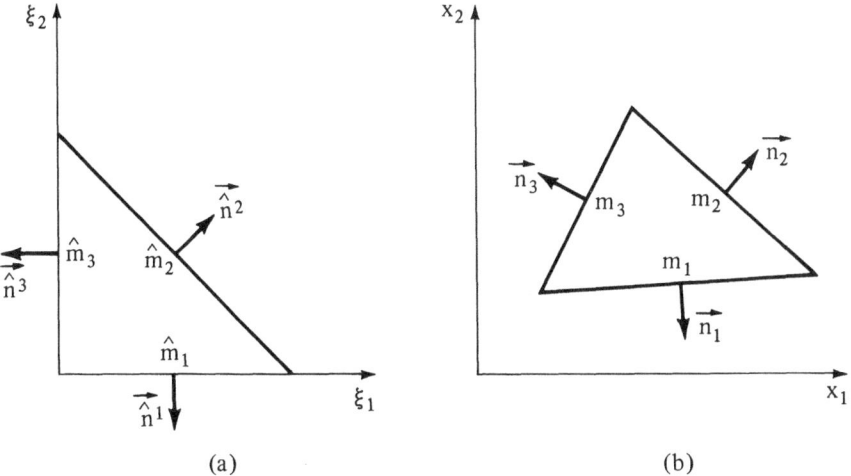

(a)                                                                    (b)

Fig. 16 (a) The reference triangle $\hat{T}$, $\hat{m}_i$: mid side nodes, $\hat{n}^i$ = unit normal vectors, $i = 1, 2, 3$; (b) Triangle $T$, $m_i$ = mid side nodes, $n^i$ unit normal vectors pointing outwards of $T$

These equations can readily be assembled from (50), (51), (53), (54).

We now state the following sufficient (this is valid for any system of the form (54)) conditions for convergence:

(i) The bilinear form $a(\cdot, \cdot)$ is coercive on $L^2$ [32]

$$a(\mathbf{q}, \mathbf{q}) \geq C_1 \int_\Omega |q|^2 \, dx \tag{55}$$

(ii) Brezzi-Babuška's condition:

$$\forall v \in M_h: \; \sup_{q \in X_h} \frac{b(\mathbf{q}, v)}{\|q\|_X} \geq C_2 \left( \int_\Omega v^2 \, dx \right)^{1/2} \tag{56}$$

with

$$\|q\|_X^2 = \int_\Omega \left( |\mathbf{q}|^2 + \left( \operatorname{div} \mathbf{q} \right)^2 \right) dx$$

Note that the constants $C_1, C_2$ should be independent of the meshsize. The meaning of (55)–(56) is as in the 1-dimensional case:

(i) For a given $u_h$, (54a) determines uniquely the discrete gradient $p_h$;
(ii) $p_h$ is a genuine gradient of $u_h$, i.e., $p_h = 0$ together with boundary conditions imply $u_h = 0$.

For the proof of (56), see e.g. Raviart-Thomas (1977a) or Thomas (1977).

---

[32] In a more general case, $a(\cdot, \cdot)$ should be coercive over the space of $q$ such that $b(\mathbf{q}, v) = 0 \; \forall v$, for the norm on $X$.

**Remarks.**

(i) If the bilinear form $a(\cdot,\cdot)$ is symmetric, i.e. if $\kappa_{ij}=\kappa_{ji}$ then (54) can be reformulated as finding the saddle point of the Lagrangian functional:

$$\mathcal{L}(\mathbf{p},v)\leq\mathcal{L}(\mathbf{p},u)\leq\mathcal{L}(\mathbf{q},u),\quad\text{for all }\mathbf{q}\in X_h,\,v\in M_h$$

with

$$\mathcal{L}(\mathbf{q},v)=\tfrac{1}{2}a(\mathbf{q},\mathbf{q})+b(\mathbf{q},v)-(f,v)$$

In elasticity theory[33] the corresponding statement is known as "*Hellinger-Reisner principle*"

(ii) From a practical point of view, the main difficulty of the method is technique of solution of the linear system;

(iii) For more details and numerical results[34] on application in structural mechanics, see Brezzi et al. (1979), Pian (1971) and the references of these papers.

---

[33] with deplacement-stresses as dependent variables.

[34] For the implementation, the expression of basis functions of such finite elements, see Jaffre (1979a), O. H. El Manouzi (1979).

# 2. Upwind Finite Element Schemes

One of the major difficulties encountered in the numerical simulation of fluid flows lies in the discretisation of the advection term $\mathbf{u} \cdot \mathbf{grad}\, \theta$ ($\mathbf{u}$=velocity, which is assumed to be known in this chapter; $\theta$=advected quantity: concentration, temperature, vorticity,..., or momentum). We begin with a short description of upwind finite difference schemes, since the ideas involved are underlying some of the upwind finite element schemes.

## 2.1. Upwind Finite Differences

(Roache (1972), Richtmeyer and Morton (1967) and the bibliography therein).

It is well known to finite difference users that in advection dominated flows, the use of *centered* differences to approximate the advection term, i.e.:

$$u \frac{\partial \theta}{\partial x}_{\text{(at node }j)} \sim u_j \frac{\theta_{j+1} - \theta_{j-1}}{2\partial x} \tag{57}$$

can cause very severe non physical oscillations. To be specific, consider the one dimensional stationary heat equation with advection:

$$-\kappa \frac{d^2\theta}{dx^2} + u \frac{d\theta}{dx} = 1 \tag{58}$$

with the boundary conditions:

$$\theta(0) = \theta(1) = 0 \tag{59}$$

If $\kappa \ll u$, the solution $\theta$ has a *boundary layer* of width $O(\kappa/u)$; (see Figure 17 where the exact solution is shown for $u=1$). Consider a finite difference approximation of (58) with centered approximation of the convection term:

$$\kappa \cdot \frac{2\theta_j - \theta_{j-1} - \theta_{j+1}}{\delta x^2} + u \frac{\theta_{j+1} - \theta_{j-1}}{2\delta x} = 1 \tag{60}$$

$$\theta_0 = \theta_N = 0, \qquad \delta x = 1/N$$

Let us consider a fixed meshsize $\delta x$: as $\kappa \to 0$, in the boundary layer, the continuum problem will require $d\theta/dx$ to take arbitrary large values while $\theta$

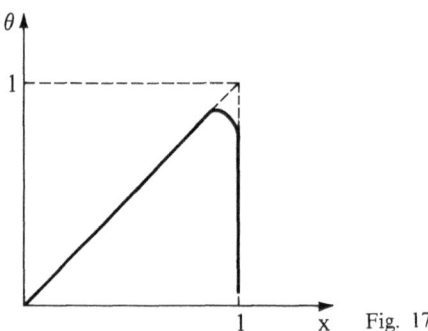

Fig. 17

remains bounded; on the other hand a fixed mesh is unable to represent arbitrary large gradients unless $\theta_j$ is itself bounded.

Therefore when the ability of the mesh to resolve $d\theta/dx$ is reduced, the imbalance in equation (58) is absorbed by the diffusion term, and spatial oscillations are induced: a careful analysis of the production of such "wiggles" is given by Roache (1972, §III-C-8, pp. 161–165). Namely, wiggles occur when:

$$\frac{u\,\delta x}{\kappa}\,(=\text{cell Reynolds number})\geq 2.$$

We note that when $\kappa$ becomes small, equation (58) degenerates into a first order equation for which a prescription of 2 Dirichlet boundary conditions is inadequate: indeed the numerical solution is "polluted" far inside the domain by the boundary condition at $x=1$.

**Remark.** If the boundary condition at $x=1$ is of Neumann type: $d\theta/dx=0$, there are no wiggles even with a centered approximation of the advection (for a discussion of this point, see Gartling (1978), Griffiths (1976)).

Remedies to numerical wiggles, in finite difference context is now well established.

Consider the *upwind* finite difference scheme (or one sided scheme):

$$\kappa\cdot\frac{2\theta_j-\theta_{j-1}-\theta_{j+1}}{\delta x^2}+u\cdot\left\{\begin{array}{ll}\dfrac{\theta_j-\theta_{j-1}}{\delta x}, & \text{if } u>0\\[2ex]\dfrac{\theta_{j+1}-\theta_j}{\delta x}, & \text{if } u<0\end{array}\right\}=1 \qquad (61)$$

Then the solution of (61) suffers no wiggles, regardless to the respective values of $\kappa, u, \delta x$. We note that the solution of (61) is practically not influenced for small $\kappa$, by the condition at the exit boundary (at $x=1$ for $u>0$): the approximation of the advection term uses exclusively information from *upstream*.

The discretisation of the advection is only first order accurate; a second order scheme can be constructed in the following way (see Roscoe (1974, 1975, 1976), Barrett (1974) for recent developments): the scheme gives the exact solution of the constant-coefficient equation; for example, the exact solution of (58) is, for constant $u$ and $\kappa$:

$$\theta(x) = u^{-1}(x - (1 - \exp(\bar{\gamma}x))/(1 - \exp\bar{\gamma})) \tag{62}$$

with $\bar{\gamma} = u/\kappa$.

Then it is easy to build a difference scheme whose solution is exact at the mesh points, namely:

$$\frac{\theta_{j+1} - (1 + \exp(\gamma))\theta_j + \exp(\gamma)\theta_{j-1}}{u^{-1}\delta x(1 - \exp\gamma)} = 1 \tag{63}$$

with

$$\boxed{\gamma = \bar{\gamma}\,\delta x = u\,\delta x/k.}$$

We can rearrange (63) as:

$$u\frac{\theta_{j+1} - \theta_{j-1}}{2\,\delta x} + \kappa\frac{\gamma}{2}\frac{1 + \exp(\gamma)}{1 - \exp(\gamma)}\left(\frac{\theta_{j+1} - 2\theta_j + \theta_{j-1}}{\delta x^2}\right) = 1 \tag{64}$$

so that the scheme adds a *numerical viscosity*:

$$\boxed{\kappa' = \kappa\left(\frac{\gamma}{2}\coth\frac{\gamma}{2} - 1\right).} \tag{65}$$

Now, if the data $\kappa$, $u$, are $x$-dependent, the scheme (63) can still be applied at each grid point and has optimal $O(\delta x^2)$ accuracy.

It will be instructive to consider another way to achieve optimal accuracy via upwinding: instead of the upstream derivative in (61) we take a combination of upstream and downstream derivatives:

$$\kappa\frac{2\theta_j - \theta_{j-1} - \theta_{j+1}}{\delta x^2} + u\cdot\left[\alpha\frac{\theta_j - \theta_{j-1}}{\delta x} + (1 - \alpha)\frac{\theta_{j+1} - \theta_j}{\delta x}\right] = 1 \tag{66}$$

where

$$1 \geq \alpha \geq 1/2, \quad \text{for } u \geq 0$$
$$0 \leq \alpha \leq 1/2, \quad \text{for } u \leq 0.$$

(Note that (61) is of this form with $\alpha = \pm 1$; $\alpha = 1/2$ corresponds to centered differencing; $|\alpha| < 1/2$ would give a "down winding" scheme and is of course unstable).

It is quite straightforward to check that scheme (66) adds a numerical viscosity equal to:

$$\kappa' = \kappa\gamma(\alpha - 1/2) \tag{67}$$

By comparison with (65) we see that optimal accuracy can be achieved with the following choice of $\alpha$:

$$\boxed{2\left(\alpha - \frac{1}{2}\right) = \coth\frac{\gamma}{2} - 2/\gamma} \tag{68}$$

that is:

$$\alpha = \frac{1}{2} + \frac{1}{2}\coth\frac{\gamma}{2} - 1/\gamma.$$

Our point here is that formula (68) is used in some upwind finite element schemes to find the "best" upwinding parameter, as will be seen in the following sections.

Indeed the standard *finite element* methods presented in chapter 1 will lead to "centered" approximations of the advection term, and numerical wiggles do occur when boundary layers are poorly represented (Figure 17). So that the *direction* of flow must be taken into account in the discretisation of advection: there are several ways of using this information, and therefore several different "families" of upwind finite element methods.

We note however that upwinding is still far from being a standard technique in finite elements: the proof of convergence is not firmly established for most of those methods, and systematic comparisons are still to come; this is a subject of active research.

Before reviewing these methods we turn to the one dimensional transport equation:

$$\frac{\partial\theta}{\partial t} + c\frac{\partial\theta}{\partial x} = 0$$

which represents a wave propagation at speed $c$: if centered finite differences (or element) are used, a perturbation at some point $x_0$ will be propagated in the wrong (upstream) direction as well as in the downstream direction; on the other hand, if upwind differencing is used, the perturbation is propagated only downstream.[35] In general upwind finite element schemes try to preserve this property, by using the information from upstream to discretize the advection.

Alternately the method of characteristics can preserve this property.

**Remark.** Upwinding is likely to be required when boundary layers are known to be required; of course alternatives to upwinding are:

(i) To remove the boundary layers by a change of boundary conditions;
(ii) To refine the mesh in the appropriate zone. (Gartling (1978)). However this is not possible in all practical problems.

---

[35] This property is called the transportive property by Roache.

## 2.2. Modified Weighted Residual (MWR)

Christie et al. (1976), Heinrich et al. (1977), Zienckiewicz (1977).

Consider a differential equation of the form:

$$\begin{cases} L\theta = f \\ + \text{boundary conditions} \end{cases} \tag{69}$$

Finite element method are particular examples of weighted residual method, that is we require that the residual $L\theta - f$ should be orthogonal to all functions $\phi$ lying in the proper space:

$$(L\theta - f, \phi) = 0, \qquad \forall \phi \tag{70}$$

(This is the ground for the weak formulation of (69)). Then $\theta$ is replaced by a finite sum:

$$\theta(x) = \sum \theta_i \phi_i(x),$$

where the $\phi_i$ are the shape functions; in the classical finite element method we restrict (70) to be true for the shape functions:

$$\sum_i \theta_i (L\phi_i, \phi_j) - (f, \phi_j) = 0, \qquad \forall j. \tag{71}$$

An alternative to (71) is to take the weighting functions $\phi$ in a set of functions different from the shape functions:

$$\sum_i \theta_i (L\phi_i, w_j) - (f, w_j) = 0, \qquad \forall j. \tag{72}$$

To be specific, let us consider again the one dimensional equation:

$$\begin{cases} -\kappa \dfrac{d^2\theta}{dx^2} + u \dfrac{d\theta}{dx} = 1 \\ \theta(1) = \theta(0) = 0 \end{cases} \tag{73}$$

Consider finite elements of degree 1: the $\phi_j$ are the roof functions (Figure 18a). To obtain the weighting function $w_j$, we add to $\phi_j$ a quadratic vanishing at the nodes:

$$N(\xi) = 3\xi(1 - \xi)$$

$$w_j(x) = \begin{cases} \phi_j(x) + \dfrac{\alpha_j}{h_j} N\left( \dfrac{x - x_{j-1}}{h_j} \right), & \text{if } x_{j-1} \le x \le x_j \\[3mm] \phi_j(x) - \dfrac{\alpha_j}{h_{j+1}} N\left( \dfrac{x_{j+1} - x}{h_{j+1}} \right), & \text{if } x_j \le x \le x_{j+1} \end{cases} \tag{74}$$

(see Figure 18b; the weighting functions $w_j$ are "sucked" back in the upstream direction).

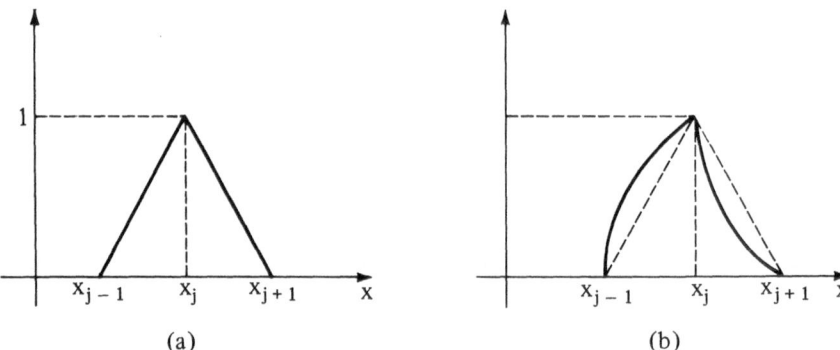

Fig. 18 (a) shape function $\phi_j$; (b) weighting function $w_j$

$\alpha_j$ is a parameter, of the same sign as $u_j = u(x_j)$, so that the weighting functions are sucked in the upwind direction.

If for simplicity we assume that $u_j$ is constant we get the scheme:

$$\kappa\left(-\frac{1}{h_j}\theta_{j-1} + \left(\frac{1}{h_j} + \frac{1}{h_{j+1}}\right)\theta_j - \frac{1}{h_{j+1}}\theta_{j+1}\right)$$

$$+ u\times\left(-\frac{1+\alpha}{2}\theta_{j-1} + \alpha\theta_j + \frac{1-\alpha}{2}\theta_{j-1}\right)$$

$$= \frac{1+\alpha}{2}h_j + \frac{1-\alpha}{2}h_{j+1}$$

$$(h_j = x_j - x_{j-1}).$$

If in addition the meshsize is uniform: $h_j \equiv \delta x$, the scheme can be written in a classical difference form:

$$\kappa\times\left(1 + \frac{\alpha u}{2\kappa}\delta x\right)\frac{2\theta_j - \theta_{j-1} - \theta_{j+1}}{\delta x^2} + u\frac{\theta_{j+1} - \theta_{j-1}}{2\delta x} = 1 \tag{75}$$

in which we identify a numerical dissipation as in (64) or (66); the numerical viscosity is: $\kappa' = \kappa \cdot \alpha\gamma/2$, $\gamma = u\delta x/\kappa$: local Reynolds number.

By comparison with the result of §2.1, we find that:

(i) The scheme is stable

either if $|\gamma| \leq 2$

or if $|\alpha| \geq 1 - 2/|\gamma|$ ("critical" value)

(ii) The *exact* solution is obtained at nodal point if:

$$\boxed{\alpha = \coth\gamma/2 - 2/\gamma} : \text{"optimal" value} \tag{76}$$

In practice for variable meshsize and velocity, $\alpha_j$ can be chosen at each point according to the law (76).

For large $\gamma$ it is quite sufficient to use, instead of the value given by (76), the critical value $1-2/\gamma_j$ which requires less computations, and which is asymptotically equivalent to (76).

The 2-*dimensional* is straightforward for quadrangles: the 2-dimensional weighting functions are just products of such 1-dimensional weighting functions (Heinrich, Huyakorn and Zienckiewicz (1977)). The numerical results given by these authors show that, even for a 2-dimensional problem, the choice of optimal $\alpha$ (76) yields a very accurate solution (it would be useful to know in which proportion this is preserved if a *distorted* mesh is used). The same authors have also generalized the method to $Q_2$ elements (Heinrich and Zienckiewicz (1977)).

As to the extension to triangular elements, they suggest to consider the triangle as a degenerate quadrangle, and to revert to the above procedure for quadrangles. An alternative and much more natural process to obtain the weighting functions could be to modify the shape functions on each triangle with the help of the bubble function: $\lambda_1(x)\lambda_2(x)\lambda_3(x)$, which vanishes along the boundary of the triangle (the $\lambda_i$ are the area coordinates).

The procedures described in this section, although effective, lead to increased computational complexity, compared to the standard Galerkin formulation:

(i) The weighting functions $w_j$ are more complicated than the shape functions, therefore higher order quadrature rules;
(ii) In a transient problem, the discretization of the temporal derivative is changed; for instance in 1 dimension, the discretization of $\partial\theta/\partial t + \partial\theta/\partial x = 0$ yields[36] with the preceding notations:

$$h_j\left(\frac{1}{6}+\frac{\alpha}{4}\right)\dot\theta_{j-1}+\left[h_j\left(\frac{1}{3}+\frac{\alpha}{4}\right)+h_{j+1}\left(\frac{1}{3}-\frac{\alpha}{4}\right)\right]\dot\theta_j$$

$$+h_{j+1}\left(\frac{1}{6}-\frac{\alpha}{4}\right)\dot\theta_j-\frac{1+\alpha}{2}\theta_{j-1}+\alpha\theta_j+\frac{1-\alpha}{2}\theta_{j+1}=0$$

($\dot\theta_j=d\theta_j/dt$: to be discretized by any appropriate scheme such as Leap Frog, Euler-backward, etc...). Note that the discretization of time derivatives is modified by this upwinding.

## 2.3. Reduced Integration of the Advection Term

Hughes (1978), Hughes et al. (1979).
Parent schemes: Dervieux (1979, unpublished), Raviart (1979).

The implementation of this successful scheme which is based on numerical quadrature techniques, is much simpler at least in the case of quadrangles; it is not subject to the previous criticisms. However its theoretical foundations

---

[36] At an interior node.

do not seem, at this moment, to be much more firmly established[37]; on the other hand its application to triangular elements seems rather unnatural.

Let us assume for the moment that we use quadrangular elements (or bricks in 3-D.); thus any element $K$ is the image of the reference element (see §1-4-3, Figure 10); we let $O^K$ be the "centre" of $K$, that is the image through the transformation of $F_K$ of point $(1/2, 1/2)$ in $\xi$ plane (in 1 dimension, $O^K$ is just the mid point of the interval).

We denote $\phi_i$ the shape functions.

Consider again the equation

$$-\kappa \Delta\theta + \mathbf{u} \cdot \mathbf{grad}\,\theta = f$$
$$\theta_{|\Gamma} = 0$$

In matrix form the finite element equations are:

$$A \cdot X = F$$

with:

$$X = \begin{bmatrix} \theta_1 \\ \vdots \\ \theta_N \end{bmatrix} \qquad \theta_h(x) = \sum_i \theta_i \phi_i(x)$$

$$F = \begin{bmatrix} f_1 \\ \vdots \\ f_N \end{bmatrix} \qquad f_i = \int_\Omega f\phi_i\,dx$$

$$A = [a_{ij}]$$

In the standard finite element method we should have:

$$a_{ij} = \sum_K a_{ij}^K,$$

where $a_{ij}^K$ is the contribution of element $K$ to the matrix:

$$a_{ij}^K = \int_K \left(\kappa\,\mathbf{grad}\,\phi_i \cdot \mathbf{grad}\,\phi_j + \phi_j(\mathbf{u} \cdot \mathbf{grad}\,\phi_i)\right) dx. \tag{77}$$

Here the definition of $a_{ij}^K$ is *modified* as follows:

$$a_{ij}^K = \left(\int_K \kappa\,\mathbf{grad}\,\phi_i \cdot \mathbf{grad}\,\phi_j\,dx\right)$$
$$+ \text{area}(K) \times \phi_i(x^K)\mathbf{u}(O^K) \cdot \mathbf{grad}\,\phi_i(x^K) \tag{78}$$

$x^K$ is some point within the element $K$: its position determines the degree of *upwinding*.

---

[37] See however Atkinson and Hughes (1977).

Let us consider in detail the one dimensional case: $K$ is some interval $x_{j-1}, x_j$ with length $h_{j-1/2} = x_j - x_{j-1}$;

$$O^K = x_{j-1/2} = (x_{j-1} + x_j)/2, \qquad x^K = x_{j-1/2} + \alpha_{j-1/2} h_{j-1/2}/2$$

$$u_{j-1/2} = u(x_{j-1/2}), \qquad \kappa_{j-1/2} = \kappa(x_{j-1/2})$$

For instance we use a 1-point Gauss formula for the integration of the diffusion term; then (78) yields at an interior node the following scheme:

$$\begin{aligned}
\theta_{j-1}&(-\kappa_{j-1/2}/h_{j-1/2} - u_{j-1/2}(1+\alpha_{j-1/2})/2) \\
&+\theta_j \times ((\kappa_{j-1/2}/h_{j-1/2} + \kappa_{j+1/2}/h_{j+1/2}) \\
&+[(1+\alpha_{j-1/2})u_{j-1/2} - (1+\alpha_{j+1/2})u_{j+1/2}]/2 \\
&+\theta_{j+1}(-\kappa_{j+1/2}/h_{j+1/2} + u_{j+1/2}(1-\alpha_{j+1/2})/2) = f_j
\end{aligned} \qquad (79)$$

Thus we obtain difference formulae very close to Zienckiewicz' method, or to finite difference formulae; indeed in the *constant coefficient case* and *uniform* mesh we get the same[38] scheme:

$$\frac{\kappa}{\delta x}(-\theta_{j-1} + 2\theta_j - \theta_{j+1}) + u\left(\frac{1+\alpha}{2}\right)(\theta_j - \theta_{j-1})$$

$$+ u\left(\frac{1-\alpha}{2}\right)(\theta_{j+1} - \theta_j) = f_j$$

So that the conclusions as to choice of $\alpha$ still holds ($\gamma = u\,\delta x/\kappa$):

$$\begin{aligned}
|\gamma| < 2: &\quad \text{stable for all } \alpha \\
\gamma > 1: &\quad \text{stable for } \alpha > 1 - 2/\gamma \\
\gamma < -2: &\quad \text{stable for } \alpha < -1 + 2/\gamma
\end{aligned} \qquad (80)$$

In all cases: $\alpha = \coth \gamma/2 - 2/\gamma$ gives optimal accuracy.

(Note that, in order to achieve the upstream discretization we evaluate the shape function at $x^K$, downstream of $O^K$ where $\mathbf{u}$ is evaluated). We turn to the two dimensional generalization and consider the bilinear isoparametric $Q_1$ element (§1-4-3 and Figures 10, 19). For any element $K$, we let $F_K$ be the bilinear transformation from the reference element $\hat{K}$ in $\xi_1, \xi_2$ plane, to $K$; to make the contribution (78) precise we only have to specify the location of the evaluation point $x^K$. We set:

$$\mathbf{e}^1 = F_K(1, 1/2) - O^K$$

$$\mathbf{e}^2 = F_K(1/2, 1) - O^K$$

$$\gamma_i = \mathbf{u}(O^K) \cdot e_i / \kappa(O^K) \qquad (i = 1, 2: \text{ element Reynolds numbers}).$$

Then we set $x^K = F_K(\xi_1, \xi_2)$ where $\xi_1$ and $\xi_2$ are computed from the values of $\gamma_1, \gamma_2$, and the one dimensional rule (80).

---

[38] For the transient equation, the schemes are different one of each other.

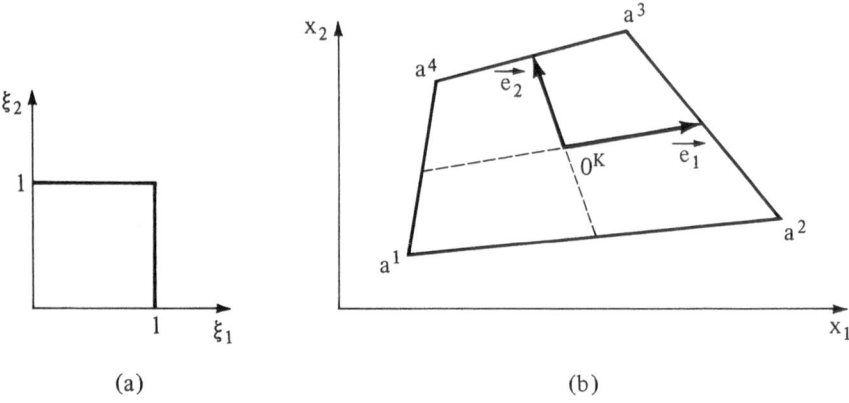

Fig. 19 (a) $\hat{K}$; (b) $K$

This procedure can be extended to higher order elements: Hughes et al. (1979) extend the scheme to the $Q_2$ element; good results are reported by these authors using rather regular meshes; it will be useful to know how the accuracy behaves when more distorted grids are used.

We have defined the scheme for quadrangles, but it can also be extended to triangles: in fact there are several ways to consider a triangle as a degenerate quadrangle (Figure 20): the choice of a fourth point ($a_4$ in Figure 20a, $a_2$ in Figure 20b) perfectly defines the mapping $F_K$, so that this upwinding procedure obviously applies. However the choice of the fourth point is somewhat arbitrary and the relation of this choice to the accuracy is not known.

A parent scheme has recently been proposed by A. Dervieux (1979, unpublished), in the framework of triangular P1 approximation; the shape functions coincide on each triangle with the barycentric coordinates.

We note $p, q, r$, the mid points of the triangle's edges (Figure 21). Then using the 3-point integration formula, the advection term in (77) becomes:

$$\int_K \phi_j (\mathbf{u} \cdot \mathbf{grad}\, \phi_i)\, dx = \frac{\text{area}(K)}{3} \sum_{m=p,q,r} \phi_j(m)\mathbf{u}(m) \cdot \mathbf{grad}\, \phi_i$$

$$\text{(grad } \phi_i \text{ is constant on the element).} \qquad (81)$$

Consider for instance the term $\phi_j(p)\mathbf{u}(p)$ in (81): $\mathbf{u}(p)$ can be decomposed along $\mathbf{pq}$ and $\mathbf{pr}$[39] as:

$$\mathbf{u}(p) = u_1(p)\mathbf{pq} + u_2(p)\mathbf{pr};$$

Then in (81), we replace $\phi_j(p)\mathbf{u}(p)$ by:

$$\phi_j(p_1)u_1\mathbf{pq} + \phi_j(p_2)u_2\mathbf{pr}$$

---

[39] Other decompositions can be tried.

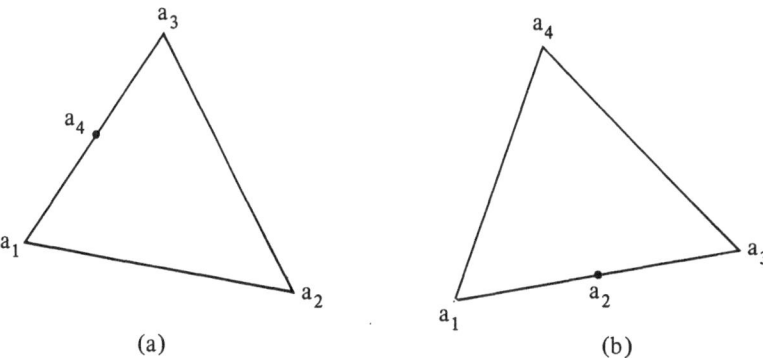

Fig. 20

where the evaluation points $p_1$ and $p_2$ are shifted (in the *downward* direction as in Hughes' method) respectively along **pq** and **pr**.

For instance:

$$p_1 = \begin{cases} p, & \text{if } u_1 \leq 0 \\ \alpha_q + (1-\alpha)p, & \text{if } u_1 \geq 0 \end{cases}$$

where $\alpha$ is a parameter of the scheme ($\alpha = 1$: full upwinding).

Note that

$$\phi_j(p_1) = \alpha\phi_j(q) + (1-\alpha)\phi_j(p)$$

and:

$$\mathbf{pq} \cdot \mathbf{grad}\,\phi_i = \phi_i(p) - \phi_i(p).$$

It is proved that, at least for $\alpha = 1$, the matrix $A = [a_{ij}]$ has the properties:

$$\begin{cases} a_{jj} > 0 \quad a_{ij} \leq 0, \quad i \neq j \\ \sum_i a_{ij} \geq 0 \end{cases} \tag{82}$$

so that the matrix $A^{-1}$ has only non negative coefficients and the scheme has the desirable property:

$$f \geq 0 \Rightarrow \theta_h \geq 0$$

(that is when the concentration or temperature should be positive, it *is*

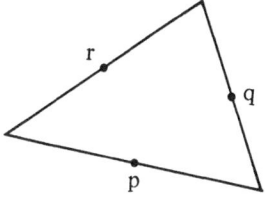

Fig. 21

positive). Furthermore it can easily be seen that the effect of the scheme is to add a numerical viscosity tensor so that the contribution $a_{ij}^K$ (77) can be replaced by:

$$\int_K \left( \kappa \delta_{rs} + \alpha_{rs}^K(u) \right) \frac{\partial \phi_i}{\partial r} \frac{\partial \phi_j}{\partial s} dx$$

where $\alpha_{rs}^K(u)$ is a positive definite pseudo viscosity tensor.

Thus the scheme comes into the framework recommended by Raviart (1979), who proposes to add just enough of numerical viscosity to make the matrix satisfy (82).

**Remark.** In the transient case, the positivity of the scheme fails unless a mass lumping approximation is made, or if the $P1$ non conforming triangle is used; this point will be discussed in §2-5 about the numerical results.

## 2.4. Computation of Directional Derivatives at the Nodes

Tabata (1977), Bristeau et al. (1979).

In such methods an approximate value of $\mathbf{u} \cdot \mathbf{grad}\, \theta$ is computed at each node, by an upstream difference-like formula, first order accurate in Tabata, second order accurate in Bristeau et al. To be specific consider the approximation by $P1$ triangles of the stationary equation:

$$-\kappa \Delta\theta + \mathbf{u} \cdot \mathbf{grad}\, \theta = f \quad \text{in } \Omega$$

$$\theta_{|\Gamma} = 0$$

The finite element equations are:

$$\theta_h(x) = \sum_i \theta_i \phi_i(x)$$

$$\kappa \left( \mathbf{grad}\, \theta_h, \mathbf{grad}\, \phi_j \right) + \left( \frac{\partial_h \theta_h}{\partial u}, \phi_j \right) = (f, \phi_j) \tag{83}$$

where $\partial_h / \partial_u$ is an approximation of the directional derivative $(\mathbf{u} \cdot \mathbf{grad})$; we use a lumping approximation for the advection term, so that (83) becomes:

$$\kappa \left( \mathbf{grad}\, \theta_h, \mathbf{grad}\, \phi_j \right) + \frac{S_j}{3} \frac{\partial_j \theta_h}{\partial u} = (f, \phi_j)$$

where $S_j$ = area of triangles surrounding node $j$ (= area of $\phi_j$'s support); and $\partial_j \theta_h / \partial u$ = approximation of $\mathbf{u} \cdot \mathbf{grad}\, \theta$ at *interior* node $j$.

In order to define this directional derivative we consider the half line starting from $\mathbf{u}$ (Figures 22 and 23). This intersects a triangle $T_j$. M. Tabata proposes to define $\partial_j \theta_h / \partial u$ as the restriction to this upstream triangle $T_j$, of $\mathbf{u} \cdot \mathbf{grad}\, \theta$ (Figure 22). Bristeau et al. look for two points $a$ and $b$ on the same

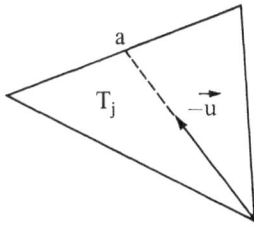

Fig. 22 Tabata's scheme: $\mathbf{u} \cdot \mathbf{grad}\,\theta|_j \sim (\theta(j) - \theta(a))|\mathbf{u}_j|$

half line from node $j$ (see Figure 23); then they define $\partial_j \theta_h / \partial u$ as the value at node $j$ of the derivative of a second order polynomial defined on the half line, coinciding with $\theta_h$ at $j, a, b$. The final formula is:

$$\frac{\partial_j \theta_j}{\partial u} = |\mathbf{u}(j)| \times \left[ \begin{array}{c} \dfrac{|ja| + |jb|}{|ja||jb|} \theta_h(j) \\[2mm] -\dfrac{|jb|}{|ja||ab|} \theta_h(a) \\[2mm] +\dfrac{|ja|}{|jb||ja|} \theta_h(b) \end{array} \right]$$

**Remarks.**

(i) When $(-\mathbf{u})$ is nearly parallel to one of the sides of the upstream triangle $T_j$, the point $b$ must be looked for further back upstream; (Figure 24) (otherwise the scheme is likely to degenerate to first order accuracy, i.e. to Tabata's scheme). See Bristeau et al. for more details.

(ii) Near the boundary it may be impossible to define a point $b$ sufficiently far upstream; in this case, the scheme reverts to Tabata's scheme.

(iii) In Tabata's scheme, the matrix has negative off diagonal elements as in previous section (Dervieux, Raviart) and $A^{-1} \geq 0$; so that the scheme enjoys the property of discrete maximum principle:

$$f \geq 0 \Rightarrow \theta_h \geq 0$$

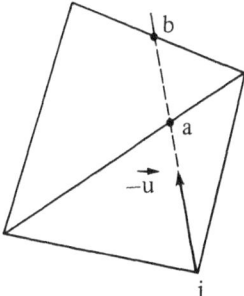

Fig. 23 Bristeau et al.'s scheme

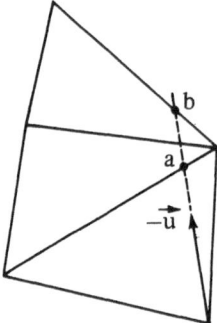

Fig. 24

In the transient case, with the mass lumping approximation for the time derivativ ε term, Tabata proves the $L^\infty$ stability of the solution.

(iv) In Bristeau et al., the band of the matrix is significantly increased compared to Tabata's scheme. On the other hand, as can be seen from the numerical results the numerical diffusion introduced by Bristeau et al. is less than in Tabata's scheme.

## 2.5. Discontinuous Finite Elements and Mixed Interpolation

Lesaint (1975), Lesaint and Raviart (1974), Jaffre (1979), Raviart (1979).

The difficulties encountered in fluid flow problems by classical finite element methods are due principally to the inability of *continuous* piecewise polynomials to represent steep gradients. Therefore it seems appropriate to approximate those steep gradients (shocks or boundary layers or "free" layers) by *discontinuities* in the approximation. Thus we consider in this section an approximation $\theta_h$ to $\theta$, which is not explicitly required to match at the interelement boundaries. Then two questions arise:

(i) How can we define "grad" $\theta_h$?
(ii) How shall we define "$\Delta$" $\theta_h$?

The second question has already been answered, since mixed interpolations can be used, which have been introduced in §1-5. We turn for the moment to the first point. To be specific consider the first order one dimensional equation

$$u\frac{d\theta}{dx}+f(x,\theta)=0, \qquad 0<x<1 \tag{84}$$

$$\theta(0)=0$$

(we assume $u(0)>0$, $u(1)>0$, so that the problem is well posed). We choose $(N+1)$ grid points: $x_0=0, x_1, x_2, \ldots, x_N=1$, which subdivide $)0,1($ into $N$

elements. We look for the discrete solution, $\theta_h$, in the class $M_k$ of functions which are:

(i) Polynomial of degree $k$ on each element;
(ii) Possibly continuous at the grid points.

For any such function $\phi$, the solution of (84) satisfies:

$$\int_{x_{j-1}}^{x_j} \left[ u \frac{d\theta}{dx} + f(x, \theta) \right] \phi \, dx = 0, \qquad j = 1, \ldots, N$$

Integrating by parts on $(x_{j-1}, x_j)$, we get:

$$\int_{x_{j-1}}^{x_j} \left[ -\theta \frac{d}{dx} (u\phi) + f(x, \theta)\phi \right] dx$$
$$+ u(x_j)\theta(x_j)\phi(x_j - 0) - u(x_{j-1})\theta(x_{j-1})\phi(x_{j-1} + 0) = 0 \qquad (85)$$

with the notations:

$\phi(x_j - 0) = $ left hand side value of $\phi$ at $x_j$

$\phi(x_{j-1} + 0) = $ right hand side value of $\phi$ at $x_{j-1}$.

The discrete equations for $\theta_h$ are obtained from (85): we write instead of $\theta(x_j)$ and $\theta(x_{j-1})$ the *upstream* values of $\theta_h$ respectively at $x_j$ and $x_{j-1}$:

$\theta_j^u = $ *upstream* value of $\theta_h$ at $x_j$

$$= \begin{cases} \theta_h(x_{j-0}), & \text{if } u(x_j) > 0 \\ \theta_h(x_j + 0), & \text{if } u(x_j) < 0. \end{cases}$$

Thus we get the discrete equations:

$$\begin{cases} \theta_h \in M_k \\ \int_{x_{j-1}}^{x_j} \left[ -\theta \frac{d}{dx} (u\phi) + f(x, \theta)\phi \right] dx \\ + u(x_j)\theta_j^u \phi(x_j - 0) - u(x_{j-1})\theta_{j-1}^u \phi(x_{j-1} + 0) = 0 \\ \qquad j = 1, 2, \ldots, N \quad \forall \phi \in X_k \\ \theta_0^u = 0 \text{ (boundary condition)}. \end{cases} \qquad (86)$$

An equivalent formulation can be written by another integration by parts:

$$\int_{x_{j-1}}^{x_j} \left[ u \frac{d\theta_h}{dx} + f(x, \theta_h) \right] \phi \, dx$$
$$+ u(x_j)\phi(x_j - 0)\left[ \theta_j^u - \theta_h(x_j - 0) \right] \qquad (86a)$$
$$- u(x_{j-1})\phi(x_{j-1} + 0)\left[ \theta_{j-1}^u - \theta_h(x_{j-1} + 0) \right] = 0$$

As an example, take $u \equiv 1$ and $k = 0$ ($\phi$ and $\theta_h$ are piecewise constant); then

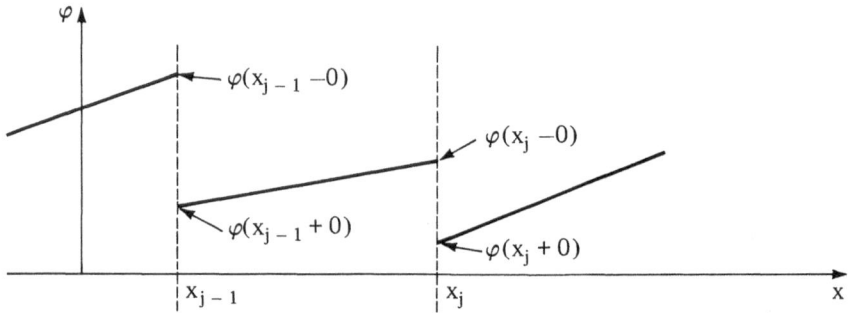

Fig. 25

$\theta_j^u = \theta_h(x_j - \dot{0})$ and (86) (or (86a)) yields:

$$\begin{cases} \int_{x_j}^{x_{j+1}} f(x, \theta_h) \, dx + \theta_{j-1/2} - \theta_{j+1/2} = 0, \qquad j = 0, \ldots, N-1 \\ \theta_{-1/2} = 0 \end{cases} \tag{87}$$

with $\theta_{j+1/2}$ = value of $\theta_h$ over $)x_j, x_{j+1}($.

Note that $u$ should be continuous all over the domain in order that (86) or (86a) makes sense.

Now we extend the scheme (86) to the *two*[40] *dimensional case*, for the equation:[41]

$$\mathbf{u} \cdot \mathbf{grad} \, \theta + \mu\theta = f \quad \text{in } \Omega$$

$$\theta = 0 \quad \text{on } \Gamma = \text{part of the boundary where the flow } \mathbf{u} \text{ is entering } \Omega, \quad (88)$$
$$\text{i.e. } \mathbf{u} \cdot \mathbf{n} \geq 0$$

We proceed as in the one dimensional case: we divide[42] $\Omega$ into elements: triangles or convex quadrangles.

Given an integer $k(=0, 1, 2, \ldots)$ we look for the approximate solution $\theta_h$ in $M_k$:

$M_k$ = set of functions which are:

Possibly discontinuous at the interelements;
Polynomial of degree $\leq k$ on each triangle;
Isoparametric $Q_k$ on each quadrangle (see Chapter 1).

For any element $K$ in the mesh, we note $\partial K$ the boundary of $K$, $\mathbf{n}_K$ the unit vector normal to $\partial K$, pointing outwards from $K$. Thus for any $\phi \in X_k$, and all

---

[40] The treatment of a three dimensional case would be similar.
[41] Problem (88) is well posed if $2\mu - \text{div}\,\mathbf{u}$ is positive and bounded away from 0.
[42] The effect of curved boundaries has not been studied.

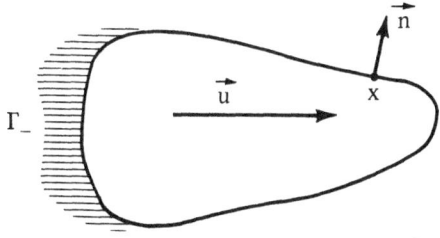

Fig. 26

$K$, (88) implies:

$$\int_K \left[ -\operatorname{div}(\mathbf{u}\phi)\theta + \mu\theta\phi - f\phi \right] dx + \int_{\partial K} (\mathbf{u}\cdot\mathbf{n}_K)\theta\phi\, ds = 0. \tag{89}$$

In order to *define the upstream* value of $\theta_h$ along $\partial K$, we set:

$\partial K$ = part of $\partial K$ where the flow $\mathbf{u}$ is entering into $K$

$$= \{ x \in \partial K \quad \text{such that: } \mathbf{u}(x)\cdot\mathbf{n}_K(x) < 0 \}$$

$$\partial K_+ = \{ x \in \partial K \quad \text{such that: } \mathbf{u}(x)\cdot\mathbf{n}_K(x) \geq 0 \}$$

(At point $x \in \partial K_-$, the flow comes from a neighbor triangle $K'$ into $K$). We set:

$\theta_h^u(x)$ = upstream value of $\theta_h$

$$= \begin{cases} \theta_{h|K'}(x) & \text{if } x \in \partial K_- \\ \theta_{h|K}(x) & \text{if } x \in \partial K_+ \end{cases}$$

Then the discrete equations are obtained from (89):

$$\int_K \left[ -\operatorname{div}(\mathbf{u}\phi)\theta_h + \mu\theta_h\phi - f\phi \right] dx$$

$$+ \int_{\partial K} (\mathbf{u}\cdot\mathbf{n}_K)\theta_h^u\phi\, dx = 0, \qquad \forall\, \phi \in M_k, \quad \forall\, K \tag{90}$$

$$\theta_h \in X_k,$$

$$\theta_h^u = 0 \quad \text{on } \partial K_- \cap \Gamma_-$$

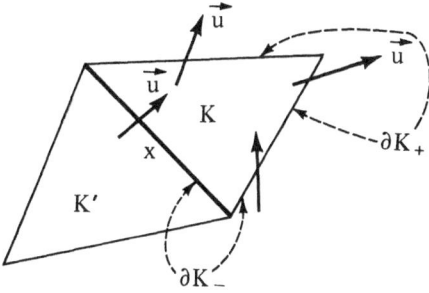

Fig. 27

With integration by parts and straightforward algebra we obtain an alternate (equivalent) formulation to (90):

$$\int_K [\mathbf{u} \cdot \mathbf{grad}\, \theta_h + \mu \theta_h - f] \phi \, dx$$
$$+ \int_{\partial K_-} (\mathbf{u} \cdot \mathbf{n}_K) [\theta_{h|K'} - \theta_{h|K}] \phi_{|K} \, ds = 0 \qquad (90a)$$
$$\forall \phi \in M_k, \quad \forall K \quad \text{in the mesh.}$$

*Error estimates:*

$$\left[ \int_\Omega |\theta_h - \theta|^2 \, dx \right]^{1/2} = O(h^k)$$

(thus the convergence holds only for $k > 0$).

This result can be improved when all elements are rectangles, in which case we have

$$\left( \int_\Omega |\theta_h - \theta|^2 \, dx \right)^{1/2} = O(h^{k+1})$$

Finally we consider the *diffusion-advection* equation:

$$-\mathrm{div}(\kappa\, \mathbf{grad}\, \theta) + \mathbf{u} \cdot \mathbf{grad}\, \theta = f, \quad \text{in } \Omega$$
$$\theta = 0 \quad \text{on } \Gamma \qquad (91)$$

As in §1-5-2, we introduce $p = \kappa\, \mathrm{grad}\, \theta$ as another variable, so that (91) may be rewritten as

$$\begin{cases} \text{a)} & -\mathrm{div}\, \mathbf{p} + \mathbf{u} \cdot \mathbf{grad}\, \theta = f \\ \text{b)} & \mathbf{p} - \kappa\, \mathrm{grad}\, \theta = 0 \\ \text{c)} & \theta = 0 \quad \text{on } \Gamma \end{cases} \qquad (92)$$

We look for a discrete solution: $\theta_h \in M_k$, $p_h \in X_k$; these approximations are not arbitrary; when the choice of $M_k$ is fixed, $X_k$ is practically determined by Brezzi's condition as in §1-5-2. Then we obtain the discrete equation just in the same manner as before:

$$\int_\Omega [\kappa^{-1} \mathbf{p}_h \cdot \mathbf{q} + \theta_h \, \mathrm{div}\, \mathbf{q}] \, dx = 0, \qquad \forall \mathbf{q} \in X_k$$

$$\int_K [-\mathrm{div}\, \mathbf{p} + \mathbf{u} \cdot \mathbf{grad}\, \theta - f] \phi \, dx$$

$$+ \int_{\partial K_-} (\mathbf{u} \cdot \mathbf{n}_K) [\theta_{h|K'} - \theta_{h|K}] \phi_{|K} \, dx = 0, \qquad \forall K, \quad \forall \phi \in M_k \qquad (93)$$

$$\left( \text{with } \theta_{h|K'} = 0 \text{ along boundary sides on } \Gamma \right)$$

**Remark.** Instead of the upstream value $\theta_h^u$, we can take, in (90), a linear combination of upstream and downstream values: $\alpha \theta_h^u + (1-\alpha) \theta_h^d$; it is a plausible (not tested) hypothesis that in the diffusion-advection problem accuracy can be improved by the choice of an "optimal" value of $\alpha$ as in Hughes' or Zienckiewicz' schemes.

*Error estimates:*

$$\left( \int_\Omega |\mathbf{p} - \mathbf{p}_h|^2 \, dx \right)^{1/2} + \left( \int_\Omega |\theta - \theta_h|^2 \, dx \right)^{1/2} = O(h^k)$$

**Remarks.**

(i) If instead of (92a) we had used the equation

$-\operatorname{div}\mathbf{p} + \mathbf{u} \cdot \mathbf{p} = f$

we would have obtained a *centered* approximation of the advection term.

(ii) In order that this upwinding scheme makes sense it is essential that: $\theta$ should be discontinuous, and the *normal* component of $\mathbf{u}$ at the interelement boundaries be continuous.

From a practical point of view, the implementation of the two dimensional method is not easy, and the computation time important, for at least three reasons:

(i) The *number of variables* is much more important than for standard finite element methods: this is for instance $O(2.5 \times NT)$ for $k=0$; $O(9 \times NT)$ for $k=1$ (for a triangular mesh, with $NT$=number of triangles);
(ii) The upwinding increases the *bandwidth* of the matrix because connexions are established between non adjacent nodes; for instance in the case $k=0$, if we have a regular net of triangles arranged in rows, with $N$ triangles per row: without upwinding the bandwidth is $O(N)$; with upwinding: $O(5N)$. This contrasts with the methods of Zienckiewicz, Hughes or Tabata which do not modify the structure of the matrix.

However a formulation of this type has lead to several practical applications:

Neutron transport equation (the original work of Lesaint);

Displacement of oil by water in porous media (numerical simulation of water flooding techniques for secondary recovery of oil). This has been implemented by J. Jaffre, G. Cohen, F. Forges; a survey is given by G. Chavent (1979).

## 2.6. The Method of Characteristics in Finite Elements

Bardos, Bercovier and Pironneau (1979), Benque, Ibler, Keraimsi and Labadie (1979).

Only recently does the method of characteristics have received some attention in a finite element context. In spite of the difficulties involved for an efficient implementation, methods of this type are very appealing, due to at least two reasons:

(i) A clear physical meaning,
(ii) Good stability properties.

Such methods are concerned with time dependent hyperbolic equations; we begin with the transport equation:

$$\frac{\partial\theta}{\partial t}+\mathbf{u}\cdot\mathbf{grad}\,\theta=0$$

$$\theta(x,t)=\theta_b(x,t),\quad\text{on }\Gamma_-=\text{part of }\Gamma\text{ where the flow enters into }\Omega. \quad(94)$$

$$\theta(x,0)=\theta_0(x)=\text{initial condition.}$$

Equation (94) merely states that the quantity $\theta$ is "transported" by the flow; i.e. the function $\theta$ is constant along the path lines (characteristic lines); these are the solutions of

$$\frac{dX}{d\tau}(\tau)=\mathbf{u} \qquad\qquad\qquad (95)$$

Assume that the solution has been computed at time $t$; then for any point $x$, the characteristic can be integrated backwards from time $t+\delta t$ to time $t$; that is we compute the solution of (95) between times $t$ and $t+\delta t$, with the final condition:

$$X(t+\delta t)=x. \qquad\qquad\qquad (95a)$$

This yields a point $x'=X(t)$, and the solution at time $(t+\delta t)$, point $x$, is simply given by:

$$\theta(x,t+\delta t)=\theta(x',t).$$

We shall assume that $\mathbf{u}$ is smooth and that $\text{div}\,\mathbf{u}=0$ (incompressible flow): otherwise several path lines can cross each other, and the method would be in difficulty.[43]

In a finite context, the solution is defined by its values at the nodes, for instance at the vertices in the $P1$ conforming approximation. Now even though $x$ is a nodal point, the foot of the characteristic arriving in $x$, i.e. the point $x'$, is not a node in general. Then two kinds of processes are feasible:

(i) Compute the value $\theta(x',t)$ by *interpolation* from the values at neighboring nodes (Benque et al. (1979)); thus the domain of dependence of node $x$ is enlarged: this is a numerical diffusion effect.
(ii) Use a *piecewise constant* approximation for $\theta$ (Bardos et al. (1979)): only one node is required per triangle. These authors prove error estimates in the maximum norm (strictly speaking in the $L^\infty$ norm):

$$\text{Sup}_{x,t}\,|\theta_h(x,t)-\theta(x,t)|=O(\delta x+\delta t)$$

---

[43] A shock, i.e. a discontinuity in $\theta$ would occur.

However this last method presents some difficulties for the approximation of a physical diffusion term: then the use of mixed interpolation is required.

In any case, the practical efficiency of such methods strongly depends on the ability of the programme to trace back the path lines, in particular upon the integration scheme for the characteristic equations (95) (for instance Benque et al. use a Runge-Kutta scheme).

We note however that the method is unconditionally *stable*: unlike the other upwinding schemes, the *numerical* domain of dependence of $x$ at time $t$ (i.e., the set of points $x$ where a value at previous time $\theta(\bar{x}, \bar{t})$ $(\bar{t}<t)$ is used to compute $\theta(x, t)$ always contains the *physical* domain.

## 2.7. Perturbation of the Advective Term: Bredif (1980)

The "centered" equations (for the 1D problem)

$$\left( K\frac{d\theta}{dx}, \frac{d\phi}{dx} \right) + \left( u\frac{d\theta}{dx}, \phi \right) = (f, \phi), \qquad \forall \phi$$

are replaced by (assuming for the moment that $K/u$ is constant)

$$\left( \frac{d\theta}{dx}, \frac{d\phi}{dx} \right) - \lambda\left( T^{\lambda}(\theta), \frac{d\phi}{dx} \right) = (\bar{f}, \phi), \qquad \forall \phi,$$

where $\lambda = u/K$, $\bar{f} = f/K$; integration by parts has been used, and $T^{\lambda}(\theta)$ is a corrector operator, to be defined element per element

$$\left( T^{\lambda}(\theta), \phi \right) = \sum_i \int_{x_i}^{x_{i+1}} T_2^{\lambda}(\theta)\phi\, dx$$

($T_i(\cdot)$ will depend on the local Peclet number $\bar{\lambda}_i = \lambda h_i$). We approximate $\theta$ by finite elements of degree $k$, i.e., on $)x_i, x_{i+1}(, \theta \in P_k$; then we require $T_i^{\lambda}(\theta) \in P_k$. Furthermore it can be proved that the accuracy will not be damaged if $T^{\lambda}(\theta)$ coincides with $\theta$ for polynomials of degree $k-1$, i.e.: $\theta \in P_{k-1} \Rightarrow T_i^{\lambda}(\theta) = \theta$.

Then $T_i^{\lambda}(\theta)$ takes the form:

$$T_i^{\lambda}(\theta) = \theta - \frac{d^k\theta}{dx^k} g_i^{\lambda}$$

where $g_i$ is some polynomial of degree $k$.

EXAMPLES.
   (i) totally upwinded formulation: operator $T_{i,up}^{\lambda}$. Let $\xi_i^j$ be $(k+1)$ points equally spaced in $)x_i, x_{i+1}(: \xi_i^j = x_i + (x_{i+1} - x_i)j/k, 0 \le j \le k$

If $\lambda > 0$   $g_i^{\lambda}(x_i) = 0$

$$g_i^{\lambda}(x_{i+1}) = C_k^{-1}$$

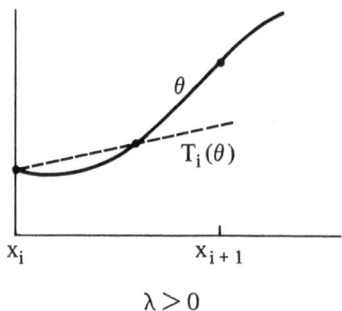

$$\lambda > 0$$

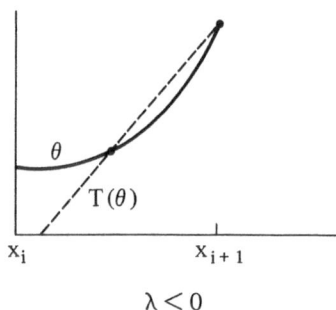

$$\lambda < 0$$

where $C_k = (k/(x_{i+1}-x_i))^k/(k-1)!$

If $\lambda < 0$   $g_i^\lambda(x_i) = (-1)^k C_k^{-1}$

$$g_i^\lambda(x_{i+1}) = 0$$

The reason of this definition is that:

if $\lambda > 0, T_i^\lambda(\theta)(\xi_i^j) = \theta(\xi_i^j)$,     $j \leq k-1$

if $\lambda < 0, T_i^\lambda(\theta)(\xi_i^j) = \theta(\xi_i^j)$,     $j \geq 1$

For instance for $k=2$:

(ii) An "optimal" operator $T_{i,\text{opt}}^\lambda$ can be defined, giving the *exact* solution for constant coefficients. Furthermore $T_{i,\text{opt}}^\lambda \to T_{i,\text{up}}^\lambda$ as $\lambda \to \pm \infty$.

This class of methods has several advantages:

Extension to variable coefficients, 2D or 3D problems, Navier-Stokes equations in straightforward, using *quadrilateral* or "*brick-shaped*" elements (the extension to simplicial elements is not really impossible, but the rules for the choice of a correction operator are not clear at the moment).

Convergence is proved.

The *bandwidth* of the resulting linear system is not increased by the introduction of the corrector.

## 2.8. Some Numerical Tests and Further Comments

The tests given in this section are intended to illustrate some of the effects of upwinding on the discretization of advection equations.

### 2.8.1. One Dimensional Stationary Advection Equation (56)

Finite elements of degree 1 were used (roof shaped basis functions) with a uniform mesh ($h = 1/80$) on the interval $)0,1($. For $\kappa = 10^{-1}$, the computed solution coincides with the exact solution as shown on Figure 17a (centered scheme).

When $\kappa = 10^{-3}$ the solution computed with the centered scheme displays, as expected, strong oscillations: Figure 17b.

The oscillations disappear when the upwinding scheme is used ($\alpha = 1$, Zienckiewcz' scheme (MWR) Figure 17c; $\alpha = .84 =$ critical value, MWR, Figure 17d).

**Remark.** In this particular case (1-D, uniform mesh, constant coefficients), Hughes' and Zienckiewicz' schemes are identical.

### 2.8.2. Two Dimensional Stationary Advection Equation

$$-\kappa \, \Delta\theta + \frac{\partial\theta}{\partial y} = 1$$

$$\theta|_\Gamma = 0$$

$$\kappa = 10^{-3}$$

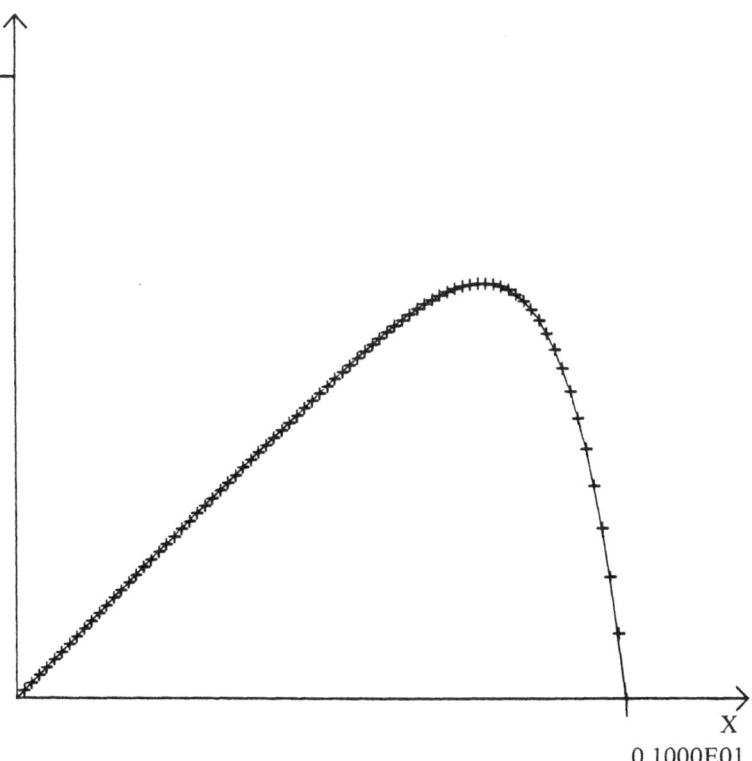

0.1000E01

+ + + + + +  : EXACT SOLUTION

——————  : COMPUTED SOLUTION

Fig. 17a $\kappa = 10^{-1}$, centered scheme

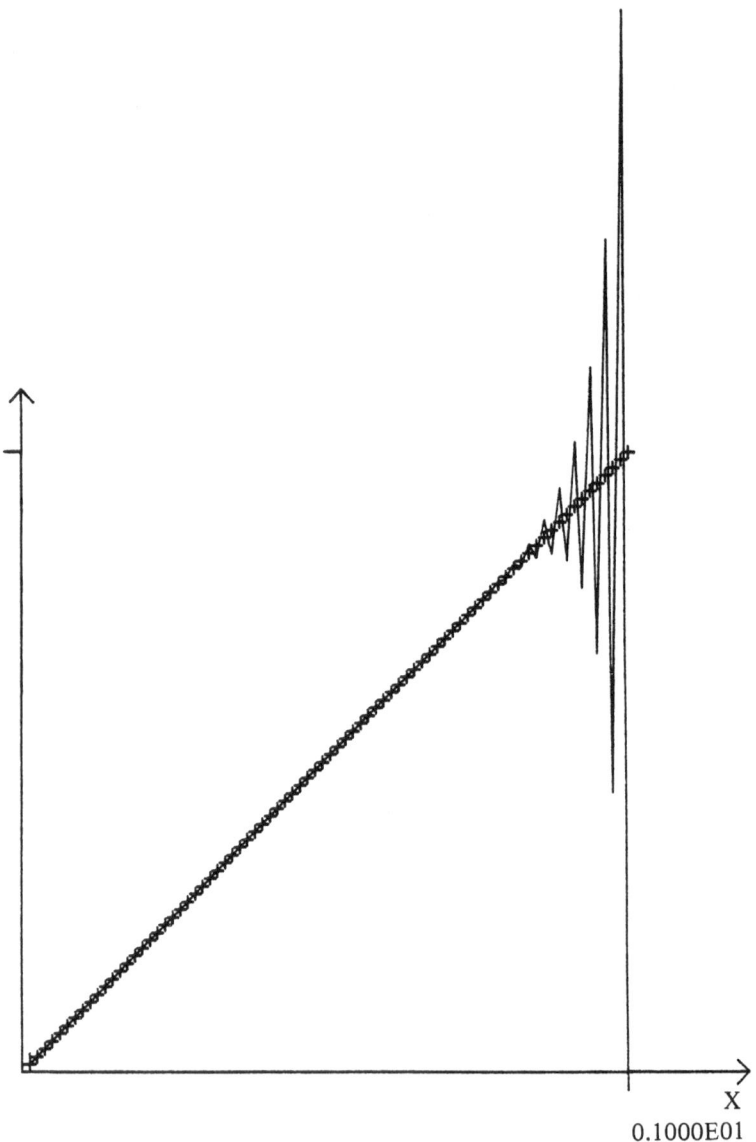

X

0.1000E01

+ + + + + + : EXACT SOLUTION

——————— : COMPUTED SOLUTION

Fig. 17b $\kappa = 10^{-3}$, centered scheme

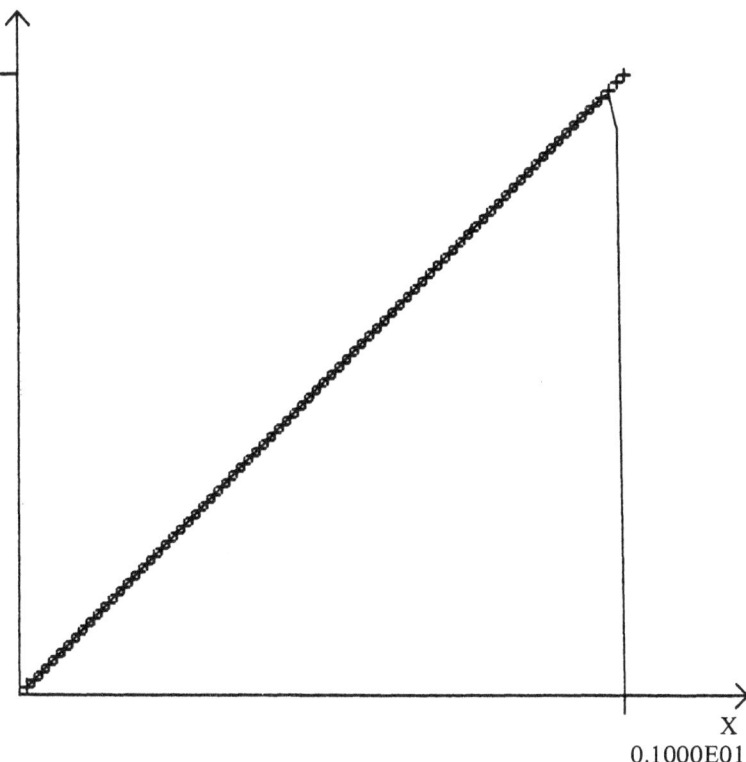

X
0.1000E01

+ + + + + +  :  EXACT SOLUTION

——————  :  COMPUTED SOLUTION

Fig. 17c $\kappa = 10^{-3}$, Zienckiewicz' scheme, $\alpha = 1$

When $\kappa$ is small the first order derivative is dominant, so that, except for $3 - y \ll 1$:

$$\theta(x, y) \sim \begin{cases} y, & \text{if } x < 1 \\ y - 1, & \text{if } x > 1 \end{cases}$$

Thus the exact solution exhibits strong gradients near $y = 3$ (boundary layer) and along $x = 1$. A uniform coarse mesh with 360 triangles has been used ($h = 1/6$); for comparison a fine mesh ($h = 1/10$, 1000 triangles).

In this case the schemes of Tabata, Bristeau et al., and (modified[44]) Dervieux are equivalent; the isolines[45] of the solution are shown on Figure

[44] Three directions are used to decompose $\mathbf{u}(p)$ (see §2-3, Figure 21):

$\mathbf{u}(p) = u_1(p)\mathbf{pq} + u_2(p)\mathbf{qr} + u_3(p)\mathbf{rp}$;

the coefficients of the decomposition are chosen so as to minimize the length $|u_1(p)| + |u_2(p)| + |u_3(p)|$, $\mathbf{u}(p)$ being fixed.

[45] I am indebted to M. O. Bristeau for these figures.

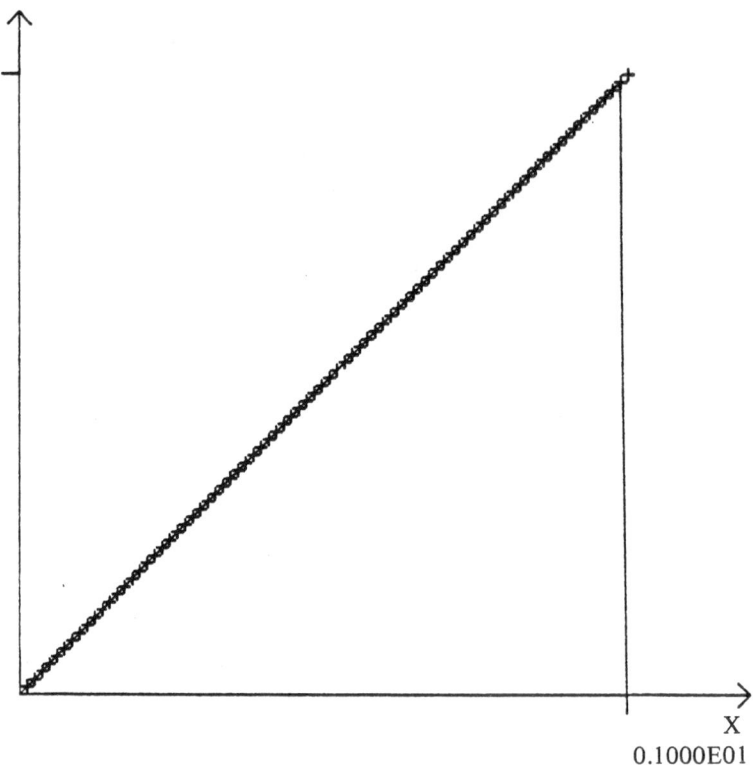

X

0.1000E01

++++++    : EXACT SOLUTION

——————  : COMPUTED SOLUTION

Fig. 17d $\kappa = 10^{-3}$, Zienckiewicz' scheme, $\alpha = .84 = 1 - 2/\gamma$

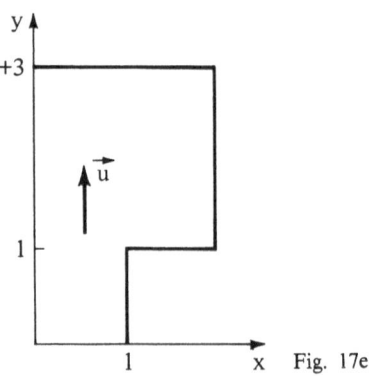

Fig. 17e

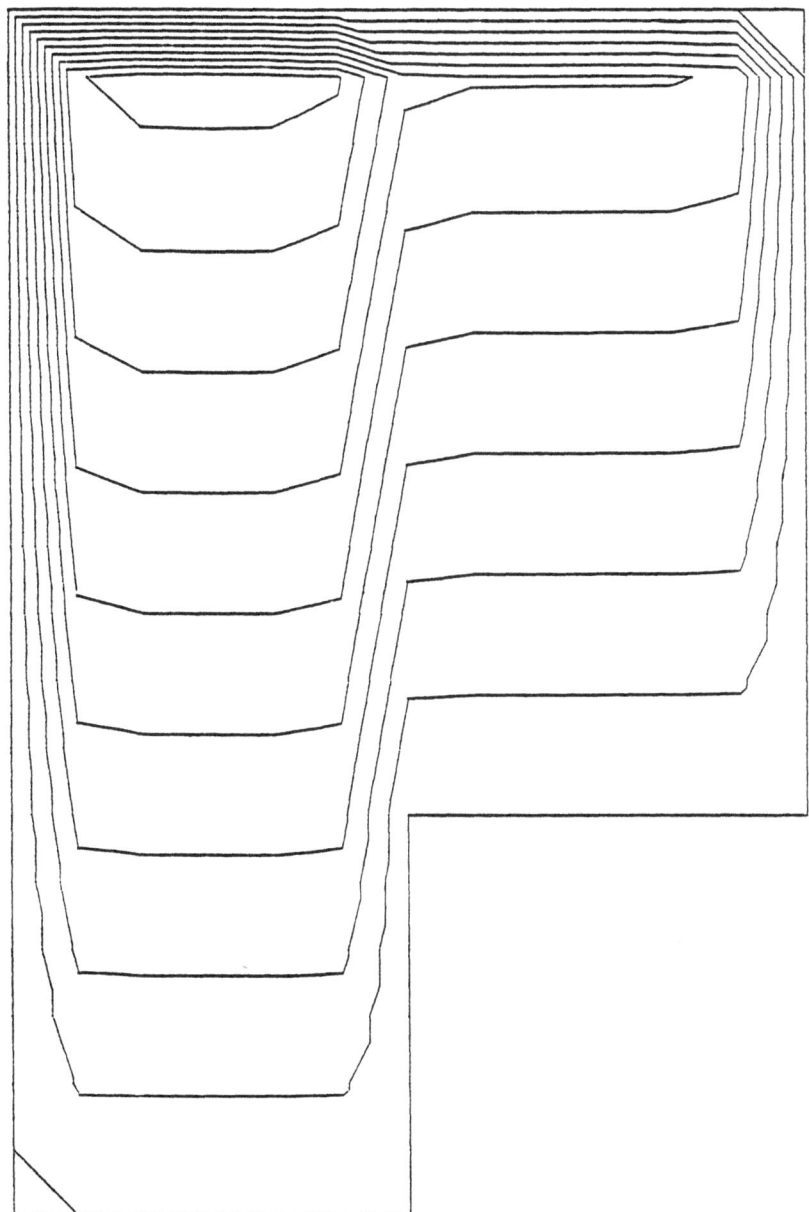

Fig. 17f $\kappa = 10^{-3}$, $P1$ conforming elements, Tabata/Bristeau et al./modified Dervieux

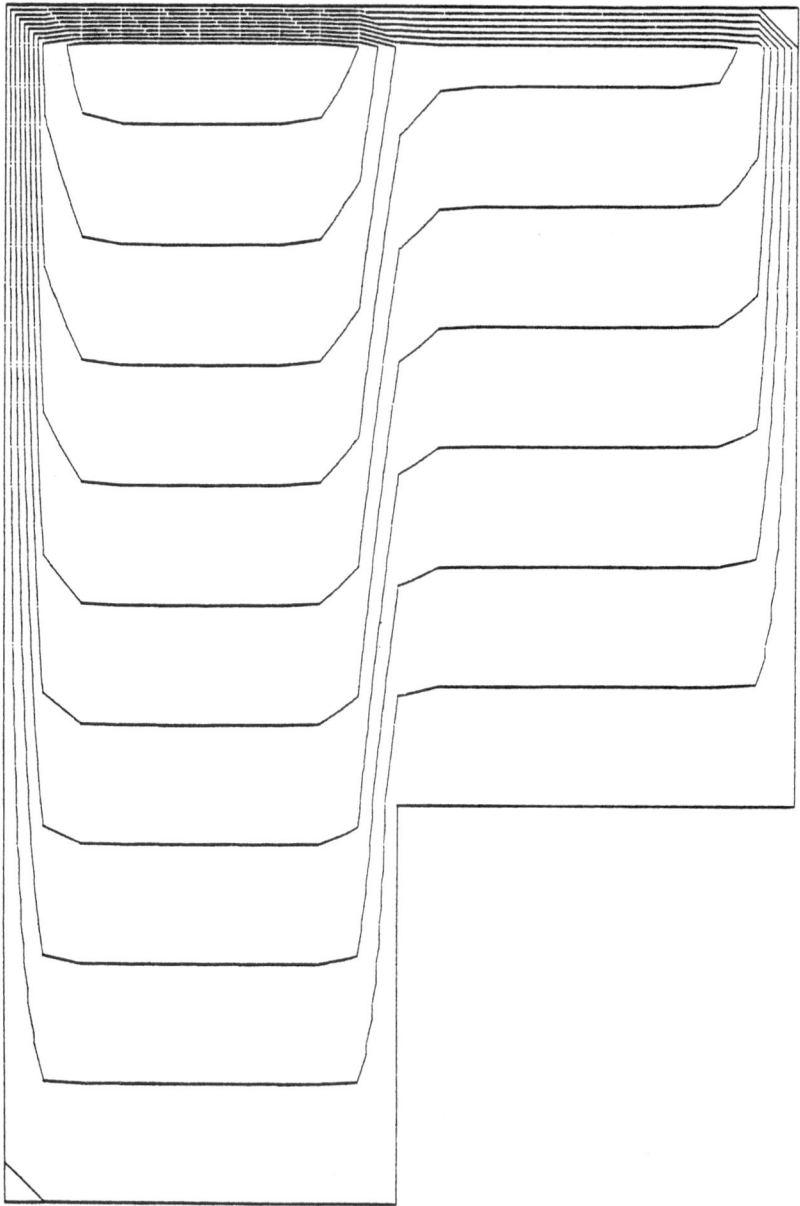

Fig. 17g $\kappa = 10^{-3}$, $P1$ conforming elements, Tabata/Bristeau et al./modified Dervieux

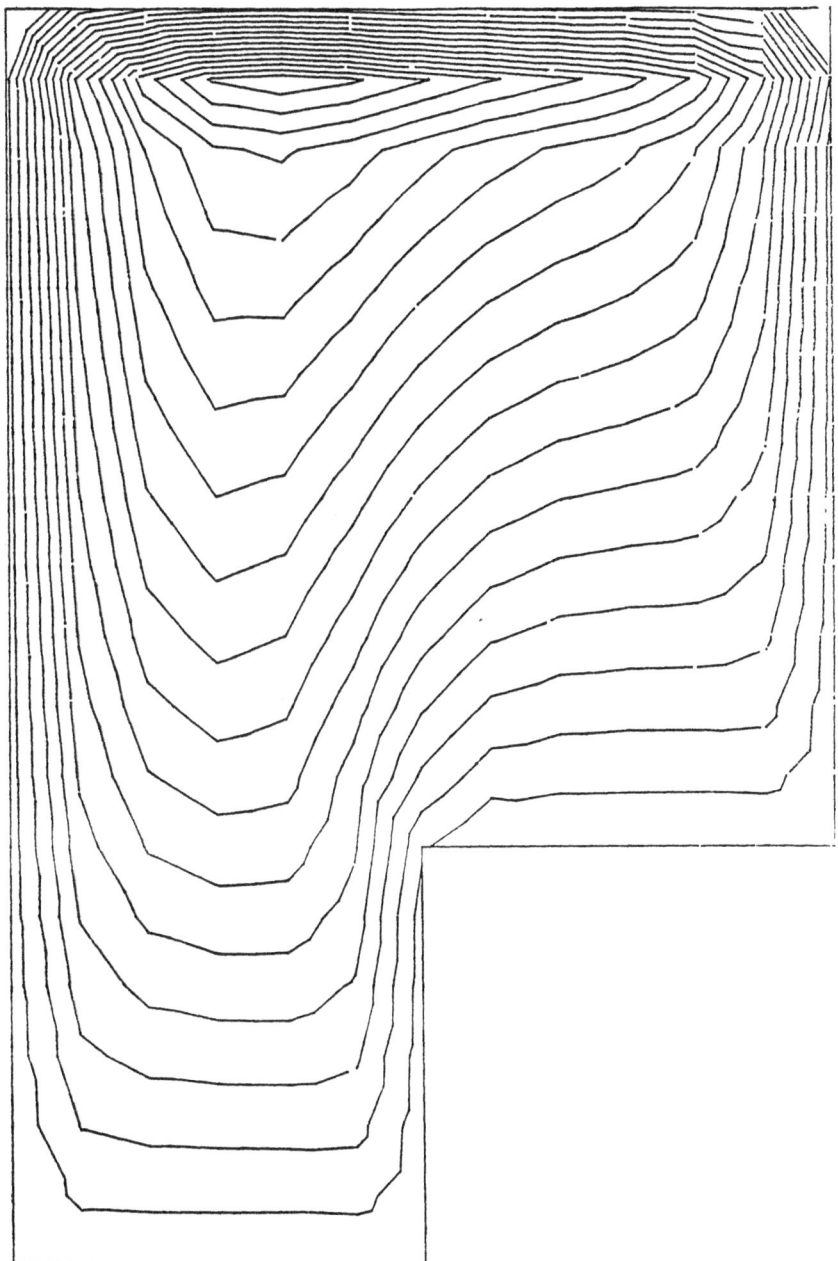

Fig. 17h $\kappa = 10^{-3}$, 360 triangles nonconforming $P1$ Dervieux's scheme

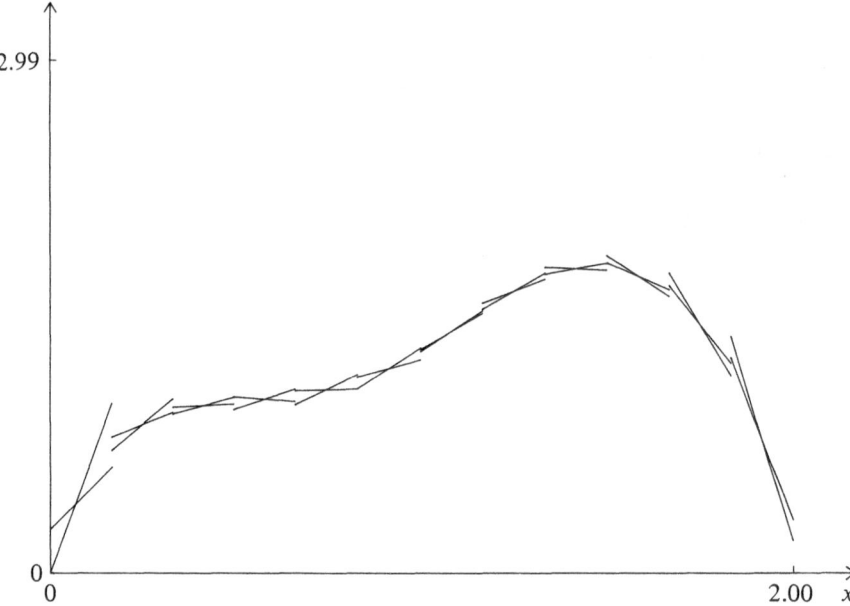

Fig. 17i $\kappa = 10^{-3}$, nonconforming $P1$, Dervieux' scheme, 360 triangles. Computed solution along $y = 2$

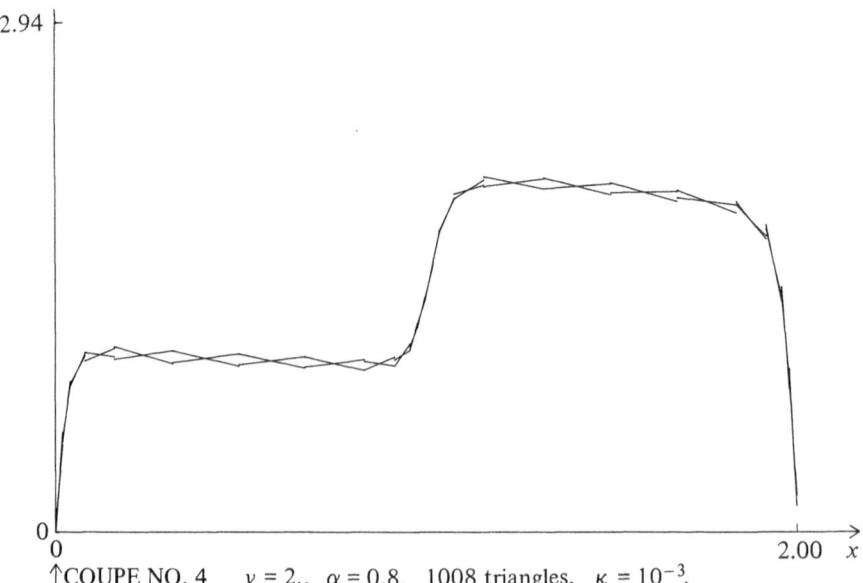

↑COUPE NO. 4     $y = 2.$, $\alpha = 0.8$, 1008 triangles, $\kappa = 10^{-3}$.

Fig. 17j $\kappa = 10^{-3}$, nonconforming $P1$, Dervieux's scheme, 1008 triangles (refined mesh near $x = 1$ and near the boundary). Computed solution along $y = 2$

17f, using the coarse mesh; with the fine the isolines are shown on Figure 17g.

When the mesh is non uniform some tests (see Bristeau et al. (1979)) indicate that the schemes of Tabata and Dervieux are more diffusive than Bristeau and Glowinski's scheme: however these tests are performed with velocities parallel to one constant direction; for a problem with rotating velocity (vortex), possibly some modifications to the original scheme might be necessary.

Dervieux's scheme has been applied with *P1 non conforming* triangles: the resulting isolines (with the coarse mesh, 360 triangles) are shown on Figure 17h: diffusive effects are more important than with the above schemes; they are further evidenced by Figure 17i and Figure 17j, displaying the variation of the computed solution along $y=2$, respectively with 360 and 1008 triangles (the last mesh is refined near the boundaries and near $x=1$).

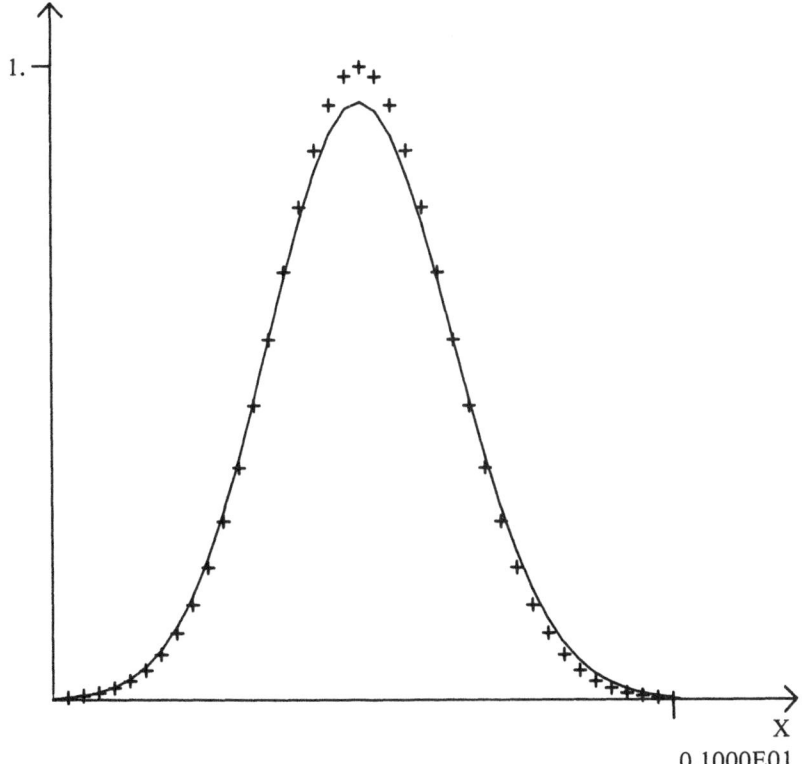

+ + + + + +  : EXACT SOLUTION

——————  : COMPUTED SOLUTION

Fig. 17k One dimensional transport equation—$h=1/40$ transport of a gaussian—solution at time $t=.5$ (centered, Zienckiewicz or Hughes schemes)

### 2.8.3. Time Dependent Advection

Let us consider the one dimensional problem:

$$\frac{\partial \theta}{\partial t} + \frac{\partial \theta}{\partial x} = 0, \qquad t > 0, \qquad 0 < x < 1$$

$$\theta(0, t) = \exp\left(-(t/2)^2\right)$$

$$\theta(x, \theta) = \exp\left(-(x/2)^2\right)$$

whose exact solution is $\theta(x, t) = \exp(-(x-t)/2)^2$, i.e. a gaussian bulb is travelling across the interval $)0, 1($. Let us take a uniform mesh ($h = 1/40$). The centered scheme yielded a reasonably accurate solution shown on Figure

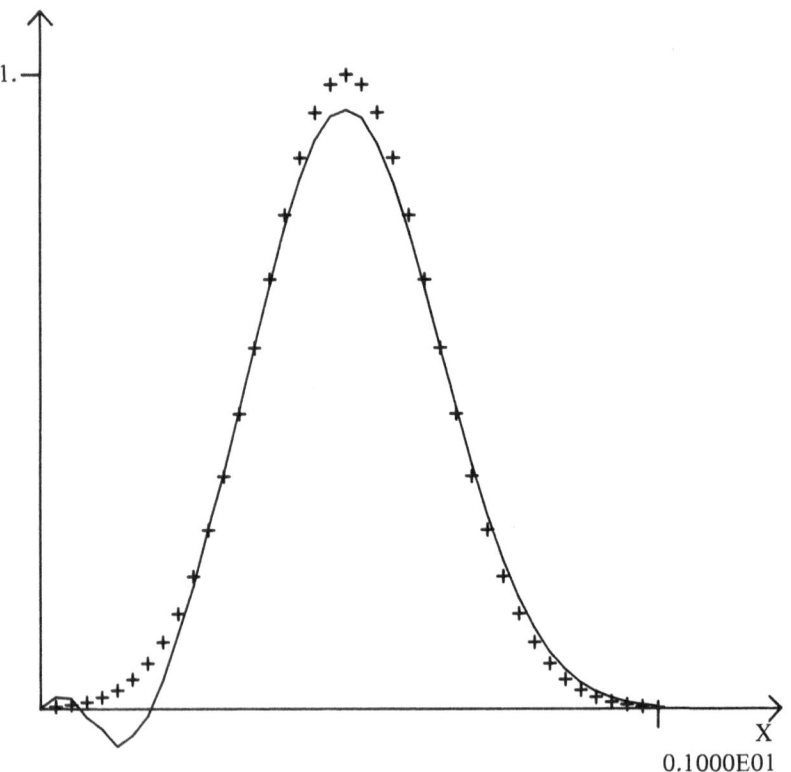

0.1000E01

+ + + + + +  : EXACT SOLUTION

────────────  : COMPUTED SOLUTION

Fig. 171 One dimensional transport equation—$h = 1/40$ centered scheme—transport of a perturbed gaussian—solution at time $t = .5$

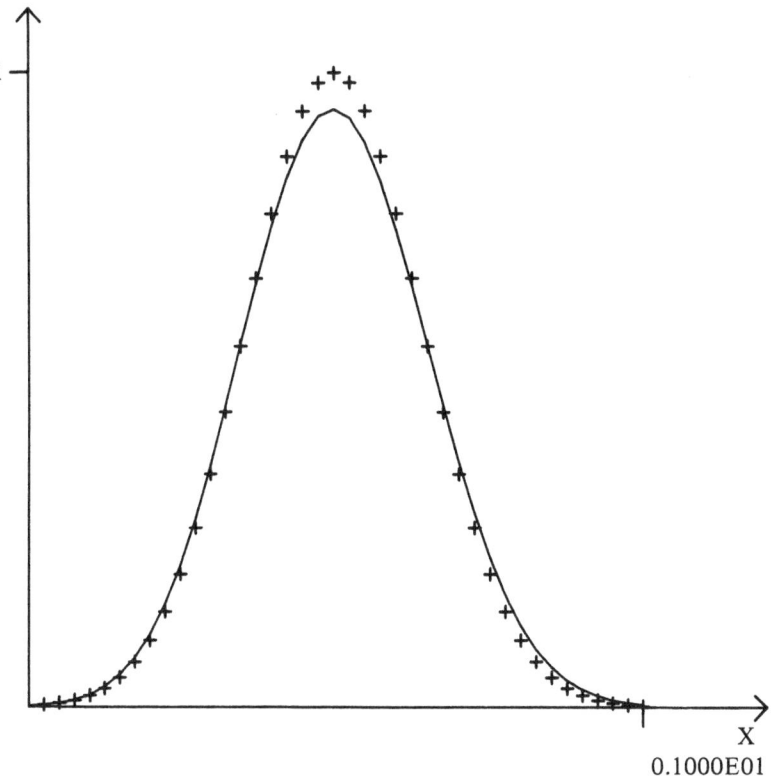

+ + + + + + : EXACT SOLUTION

————————— : COMPUTED SOLUTION

Fig. 17m One dimensional transport equation—$h=1/40$ Zienckiewicz' scheme $(\alpha=1)$—transport of an initially perturbed gaussian—solution at time $t=.5$

17k (practically the same solution is obtained using Zienckiewicz' and Hughes' schemes).

Now let us perturb the initial gaussian (initially the last right most mesh points are forced to: .4, $-.6, 1$ (from left to right: $\theta(1,0)=1$). For the exact solution, the perturbation is drifted away so that after some time the solution is that of the unperturbed problem. The computed solutions, respectively with the centered scheme, Zienckiewicz' scheme $(\alpha=1)$ Hughes' scheme $(\alpha=1)$, are shown on Figures 17l, 17m, and 17n.

As expected, with the centered scheme, the effect of the perturbation is felt upstream; that is the perturbation also propagated in the wrong direction (Figure 17l). With Zienckiewicz's scheme the perturbation is correctly drifted

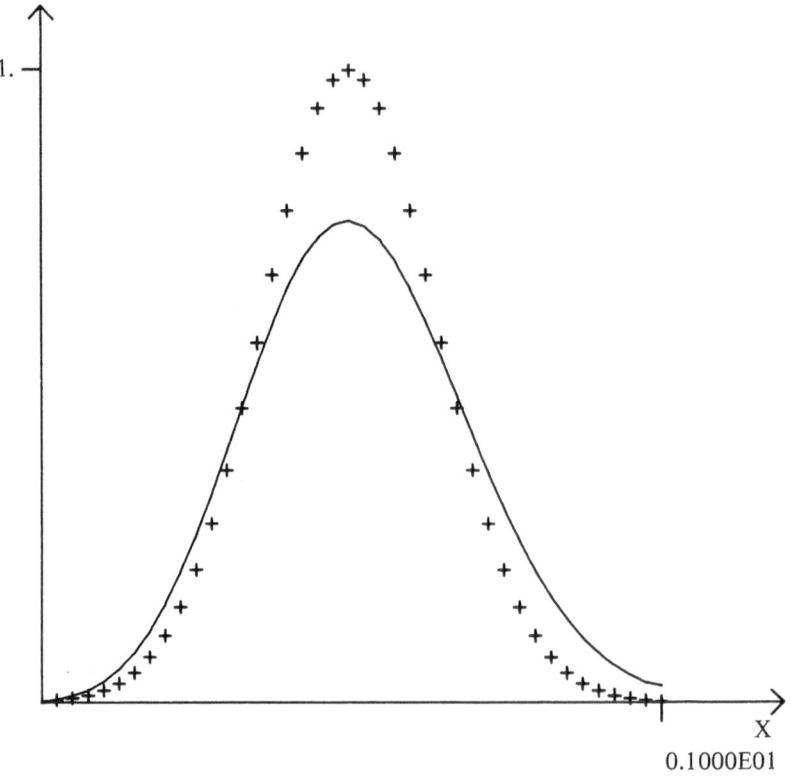

0.1000E01

+ + + + + +   : EXACT SOLUTION

——————   : COMPUTED SOLUTION

Fig. 17n One dimensional transport equation—$h = 1/40$ Hughes' scheme ($\alpha = 1$)—transport of an initially perturbed gaussian—solution at time $t = .5$

away (Figure 17m), while with Hughes' scheme the influence of the perturbation is again felt upstream (Figure 17n).

**Remark.** These calculations used a time step $\delta t = .005$ and implicit time matching scheme.

Our point[46] with these elementary tests is the following: when one wants to model fluid flows (in 2 or 3 dimensions), it is essential that transport phenomena should be correctly represented; in particular false propagation of perturbations (which is likely to occur with a centered scheme), may generate instabilities.

---

[46] The point is not to draw definite conclusions as to the relative merits of different schemes, by lack of numerical experiments.

**Remarks.**

(i) Gresho et al. (1976) displayed calculations of the propagation of a 2 dimensional gaussian in a uniformly rotating velocity field; their results tended to demonstrate the disastrous effects of *mass lumping*[47] (using $P2$ or $P1$ triangles); unfortunately little information was given as to the time marching scheme (they used a centered approximation of advection).

(ii) Benque et al. (1979) obtained qualitatively good results with their method of characteristics, on the same problem and using a $31 \times 31$ mesh, the distorsion observed after one revolution is small.

(iii) In the transient case, Tabata's scheme is proved to be stable (for the norm $L^\infty$) provided the mass lumping approximation is done, and an implicit time marching scheme is used.

---

[47] Numerical integration of time derivative term, making the corresponding matrix diagonal.

# 3. Numerical Solution of Stokes Equations

## 3.1. Introduction

A good solver of Navier-Stokes equations should at least be able to solve the Stokes equations (indeed some iterative procedures to solve Navier-Stokes equations, solve in fact a succession of Stokes problems). Therefore we are concerned in this section with Stokes equations, governing incompressible flows of newtonian fluids at negligibly small Reynolds numbers; it will be sufficient for our purpose to consider the stationary equation:

a) $\quad \sigma_{ij,j} + f_i = 0$

b) $\quad \sigma_{ij} = -p\delta_{ij} + \nu(u_{i,j} + u_{j,i})$

c) $\quad u_{i,i} = 0$ $\hfill (96)$

d) $\quad \mathbf{u} = 0$ on $\Gamma$: homogeneous boundary conditions[48]

with $\mathbf{f} = (f_1, f_2)$ (or $(f_1, f_2, f_3)$) = body force

$\qquad p$ = pressure

$\qquad \mathbf{u} = (u_1, u_2)$, or $u_1, u_2, u_3$ = velocity

$\qquad \nu$ = (constant) viscosity

$\qquad \sigma_{ij}$ = stress tensor

Using the incompressibility condition $\operatorname{div}\mathbf{u} \equiv u_{i,i} = 0$, Stokes equation can also be written as:

$$-\nu\Delta\mathbf{u} = \mathbf{f} - \operatorname{grad} p$$

$$\operatorname{div}\mathbf{u} = 0$$

$$\mathbf{u} = 0 \quad \text{on } \Gamma \hfill (97)$$

By the same standard arguments as in Chapter 1, we can establish the following weak formulations of (97):

---

[48] Other types of boundary conditions will be considered later.

Find $\mathbf{u}$ and $p$ such that[49]:

$$\nu \int_\Omega \mathbf{grad}\,\mathbf{u} \cdot \mathbf{grad}\,\mathbf{v}\,dx - \int_\Omega p\,\mathrm{div}\,\mathbf{v}\,dx = \int_\Omega \mathbf{f} \cdot \mathbf{v}\,dx$$

$$\forall \mathbf{v},{}^2 \quad \text{such that } \mathbf{v}=0 \quad \text{on } \Gamma \tag{98}$$

$$\int_\Omega q \cdot \mathrm{div}\,\mathbf{u}\,dx=0, \qquad \forall q\,(\Leftrightarrow \mathrm{div}\,\mathbf{u}=0)$$

$$\mathbf{u}=\mathbf{0} \quad \text{on } \Gamma$$

Taking in (98) functions $\mathbf{v}$ satisfying $\mathrm{div}\,\mathbf{v}=0$, we obtain:

Find $\mathbf{u} \in V$ such that:

$$\nu \int_\Omega \mathbf{grad}\,\mathbf{u} \cdot \mathbf{grad}\,\mathbf{v}\,dx = \int_\Omega \mathbf{f} \cdot \mathbf{v}\,dx, \qquad \forall \mathbf{v} \in V$$

$V$ is the class of functions[50] $\mathbf{v}$ satisfying $\tag{99}$

$$\mathbf{v}=0, \quad \text{on } \Gamma, \quad \mathrm{div}\,\mathbf{v}=0$$

(98) and (99) are equivalent weak formulations of Stokes equations (97).

At this stage we want to emphasize two points:

Equation (99) is equivalent to a *constrained optimization* problem:

$$\underset{\mathbf{v}\in V}{\mathrm{Min}}\; \frac{\nu}{2} \int_\Omega |\mathbf{grad}\,\mathbf{v}|^2\,dx - \int_\Omega \mathbf{f} \cdot \mathbf{v}\,dx. \tag{100}$$

The incompressibility condition (which is implied by "$\mathbf{v}\in V$") plays the role of a *linear constraint* in this context.

From (98), (99), (100) we can view the pressure as a *Lagrange multiplyer* associated to this linear constraint; indeed (98) are also the equilibrium equations of the saddle point problem:

$$\underset{\mathbf{v}\in H_0^1 \times H_0^1}{\mathrm{Min}}\; \underset{q}{\mathrm{Max}} \int_\Omega \left( \frac{\nu}{2}|\mathbf{grad}\,v|^2 + q\,\mathrm{div}\,\mathbf{v} - \mathbf{f}\mathbf{v} \right) dx \tag{101}$$

These two facts are of primary importance for the construction of the approximation: if the discrete incompressibility condition is of the form $\mathrm{div}_h\,\mathbf{u}_h =0$, it should be satisfied by non trivial velocity so that the discrete optimisation problem, analogue to (100), is not over constrained.

Next the discrete pressure should belong to the same class as $\mathrm{div}_h\,\mathbf{u}_h$. We make these ideas more precise with the help of the following simple example (which by the way does *not* yield a correct approximation of Stokes flows!).

Approximate $\mathbf{u}$ by the *Pl conforming* triangle (courant's triangle, §1-4-1); then the discrete velocity $\mathbf{u}_h$ is *continuous* all over the domain, and piecewise polynomial of degree $\leq 1$. We approximate the incompressibility condition

---

[49] $\mathbf{u}\in H_0^1(\Omega)\times H_0^1(\Omega), \quad p\in L^2(\Omega).$

[50] $\mathbf{v}\in H_0^1(\Omega)\times H_0^1(\Omega).$

by: $\operatorname{div}\mathbf{u}_h = 0$; this is 1 constraint per triangle, so that the discrete pressure is piecewise *constant*.

Now we count the degrees of freedom of the discrete velocity (keeping in mind the boundary conditions): this is $2\times$number of interior vertices; and this is inferior[51] to the number of linear constraints=number of triangles.

Thus, except for exceptionally regular meshes–for which some constraints can be linearly dependent–the only admissible $\mathbf{u}_h$ satisfying the boundary conditions and the incompressibility constraint, is the trivial solution $\mathbf{u}_h \equiv 0$ (Fortin (1972)).

This elementary example shows that the choices of the approximation of $\mathbf{u}$ and $p$ are not arbitrary.

We note $\mathbf{u}_h \in V_h$, $p_h \in X_h$ the approximations of $\mathbf{u}$ and $p$; the discrete analogue of (98) takes the form:

$$
\begin{cases}
\displaystyle\sum_T \int_T \nu\,\mathbf{grad}\,\mathbf{u}_h\,\mathbf{grad}\,\mathbf{v}_h\,dx - \sum_T \int_T p h\,(\operatorname{div}\mathbf{v}_h)\,dx \\
\qquad = \displaystyle\int_\Omega \mathbf{f}\cdot\mathbf{v}_h\,dx, \qquad \forall\mathbf{v}_h \in V_h \\
\displaystyle\sum_T \int_T q_h\operatorname{div}\mathbf{u}_h\,dx = 0, \qquad \forall q_h \in X_h
\end{cases}
\tag{102}
$$

Since the pressure is defined up to an additive constant, this can be fixed by assuming e.g.:

$$
\int_\Omega q_h\,dx = 0, \qquad \forall q_h \in X_h
\tag{103}
$$

Now the condition to be satisfied for convergence has the form: $\forall q_h \in X_h$,

$$
\sup_{\mathbf{v}_h \in V_h} \frac{\displaystyle\sum_T \int_T q_h\operatorname{div} v_h\,dx}{\left(\displaystyle\sum_T \int_T |\operatorname{grad} v_h|^2\right)^{1/2}} \geq c\left(\int_\Omega q_h^2\,dx\right)^{1/2}
\tag{104}
$$

where the constant $c$ is independent of the mesh. Condition (104) is satisfied if for any $q_h \in X_h$, we can prove the existence of $\mathbf{v}_h \in V_h$ such that:

$$
\sum_T \int_T (q_h - \operatorname{div}\mathbf{v}_h)\mu_h\,dx = 0, \qquad \forall\mu_h \in X_h
$$

$$
\left(\sum_T \int_T |\operatorname{grad} v_h|^2\,dx\right)^{1/2} \leq c\left(\int_\Omega q_h^2\,dx\right)^{1/2}
\tag{104a}
$$

---

[51] For instance if we take a regular $N\times N$ mesh, made of $N^2$ rectangles, ent into $2N^2$ triangles, the number of interior vertices is $(N-1)^2$; so that: number of degrees of freedom$=2\times(N-1)^2$ $<2\times N^2$=number of constraints.

We shall see further below some examples where these conditions are satisfied; for the verification of (104), (104a) and proofs of convergence (which are not simple matters) we refer to Temam (1977), Girault and Raviart (1979), Crouzeix and Raviart (1973), Fortin (1972). However the implication of condition (104) can be clearly understood: we note that in (102) the pressure gradient is discretized as:

$$\sum_T \int_T p_h \operatorname{div} v_h \, dx;$$

then if the left hand side in (104) vanishes for $p_h$ it means that the approximate gradient of $p_h$ vanishes, and (104) implies that $p_h$ also vanishes. This with (103) means that the pressure gradient in Stokes equation is properly approximated.

**Remarks.**

(i) Conditions of type (104) are encountered in the approximation of any saddle point problem (see §1-5 of this course and Brezzi (1974));

(ii) Formulation (98) is well adapted to a discontinuous approximation of the pressure as described in §3-2; in 3-3 a different formulation is used which allows *continuous* approximation of $p$;

(iii) Other types of boundary conditions can be considered, for instance:

$$\begin{cases} \sigma_{ij} n_j = \tau_i = \text{given normal stress, on } \Gamma_1 \quad (\text{part of } \Gamma) \\ u = 0 \quad \text{on } \Gamma_0 = \Gamma - \Gamma_1 \end{cases}$$

Then the proper class where $u$ is to be looked for is the space $W$ of functions $v$ vanishing on $\Gamma_0$ and the corresponding formulation is:

$$\int_\Omega \nu (u_{i,j} + u_{j,i})(v_{i,j} + v_{j,i}) \, dx - \int_\Omega p \operatorname{div} u \, dx$$

$$= \int_\Omega f_i v_i \, dx + \int_{\Gamma_1} \tau_i v_i \, ds(x), \qquad \forall v \in W$$

## 3.2 Velocity–Pressure Formulations: Discontinuous Approximations of the Pressure

The basic weak formulation is (98); the discrete pressure is discontinuous at the interelement boundaries; the interest of such approximations (over a continuous approximation of $p$ as in the next section, §3-3) is two fold:

(i) Probably[52] a better approximation of *mass conservation*;
(ii) The possibility of an easy *elimination* of the pressure, thus reducing the number of equations: this is the *penalty* technique (§3-2-3).

---

[52] This point still needs to be demonstrated by experiments.

### 3.2.1. $u_h$: P1 Nonconforming Triangle (§1-4-5); $p_h$; Piecewise Constant

This is the simplest of all finite element methods for Stokes equations; its motivation is clearly seen from the example given in the introduction: if we relax the continuity[53] requirements on $\mathbf{u}_h$, there will be more degrees of freedom, allowing the satisfaction of the incompressibility constraint.

The discrete divergence of $\mathbf{u}_h$ has the same degrees of freedom as the discrete pressure, i.e. one per triangle (as its should be since $p_h$ is the Lagrange multiplyer associated to the constraint $\mathrm{div}_h u_h = 0$). The discrete equations are readily obtained from (98):

$$\sum_T {}^{54} \int_T \nu \,\mathrm{grad}\,\mathbf{u}_h \,\mathrm{grad}\,\mathbf{v}_h \, dx - \sum_T \int_T p_h \,\mathrm{div}\,\mathbf{v}_h \, dx = \int_\Omega \mathbf{f}\mathbf{v}_h \, dx$$

$$\forall \, \mathbf{v}_h \text{ (vanishing at the boundary nodes)} \tag{105}$$

$$\mathrm{div}\,\mathbf{u}_h = 0, \quad \text{on each triangle } T$$

$$u_h = 0, \quad \text{at the boundary nodes.}$$

Taking successively $\mathbf{v}_h = (w_m, 0)$ and $(0, w_m)$ where the $w_m$ $(m \notin \Gamma)$ are the shape functions introduced in §1-4-5, we get the equations:

$$\mathbf{u}_h = \sum_{q \notin \Gamma} \begin{pmatrix} u_{1q} \\ u_{2q} \end{pmatrix} w_q$$

$$\sum_{q \notin \Gamma} u_{1q} \sum_T \mathrm{area}(T) \nu \,\mathrm{grad}\, w_{q|T} \cdot \mathrm{grad}\, w_{m|T}$$

$$- \sum_{q \notin \Gamma} \sum_T \mathrm{area}(T) p_h(T) \frac{\partial w_m}{\partial x}\bigg|_T = \int_\Omega f_q w_m \, dx$$

(and a similar equation in $u_{2q}$).

*Error estimates:* Crouzeix and Raviart (1973).

$$|u - u_h|_{1,h} \equiv \left( \sum_T \int_T |\mathrm{grad}(u - u_h)|^2 \, dx \right)^{1/2} = O(h)$$

$$|p - p_h|_0 = \int_\Omega (p - p_h)^2 \, dx = O(h), \quad \text{provided } \int_\Omega p \, dx = \int_\Omega p_h \, dx$$

The *numerical solution* of the system of linear equations (105) will be examined §3-2-5 (together with the solution of the equations obtained with similar methods).

---

[53] Along a triangle side, $\mathbf{u}_h$ is continuous at the mid point only.
[54] Summation on the triangles of the mesh.

Now, our point is just to note that with this approximation, the incompressibility condition can be easily taken into account in the *2-dimensional* case by the construction of a *zero divergence basis* (Crouzeix (1976)), so that the number of variables is reduced (the pressure is eliminated). The construction of the basis starts from the following remark: the incompressibility condition:

$$\text{div } u_h = 0 \quad \text{on each triangle } T$$

can also be written since div $u_h$ is constant per triangle:

$$\int_{\partial T} \mathbf{u}_h \cdot \mathbf{n}_T ds = 0, \tag{106}$$

(by the Gauss divergence theorem; $\mathbf{n}_T =$ unit normal vector to the boundary $\partial T$ of triangle $T$, pointing outwards of $T$). Since we use the nonconforming $P1$ triangle to define $\mathbf{u}_h$, $\mathbf{u}_h$ is determined by its values at the mid side nodes; furthermore, with the notations of Figure 28b and since $\mathbf{u}_h$ is linear:

$$\int_{aa'} \mathbf{u}_h \cdot \mathbf{n}' ds = |aa'| \mathbf{u}_h(m') \cdot \mathbf{n}' \tag{107}$$

From (106) and (107) the tangential components $\mathbf{u}_h(m') \times \mathbf{n}'$ do not play any role in the satisfaction of the incompressibility; we just have to adjust the normal components $\mathbf{u}_h(m') \cdot \mathbf{n}'$ so as to satisfy (105) on each triangle. From these considerations we define two kinds of basis functions:

*First kind:* $\mathbf{w}_m$ associated to a mid side node:

$$\mathbf{w}_m(x) = \frac{\mathbf{aa'}}{|aa'|^2} w_m(x) \,^{55}; \text{ (Figure 28a)}.$$

*Second kind:* $\mathbf{w}_a$, associated to any vertex $a$

$$\mathbf{w}_a(x) = -\frac{\mathbf{n'}}{|aa'|} w_{m'}(x) + \frac{\mathbf{n''}}{|aa''|} w_{m''}(x); \text{ (Figure 28b)}.$$

We now can develop $\mathbf{u}_h$ as:

$$\mathbf{u}_h(x) = \sum_a U_a \mathbf{w}_a(x) + \sum_m U_m \mathbf{w}_m(x) \tag{108}$$

For a given $\mathbf{u}_h$, it is an easy matter to write the equations to be satisfied by the $U_a$ and $U_m$.

We must make a distinction between several cases on geometry and boundary conditions:

*Case 1:* no obstacle in the flow, homogeneous boundary condition $\mathbf{u} = 0$; the summation in (106) is extended to all interior nodes.

---

[55] Remember that $w_m$ is the *scalar* function equal to 1 at node $m$, to 0 at all other nodes (§1-4-5).

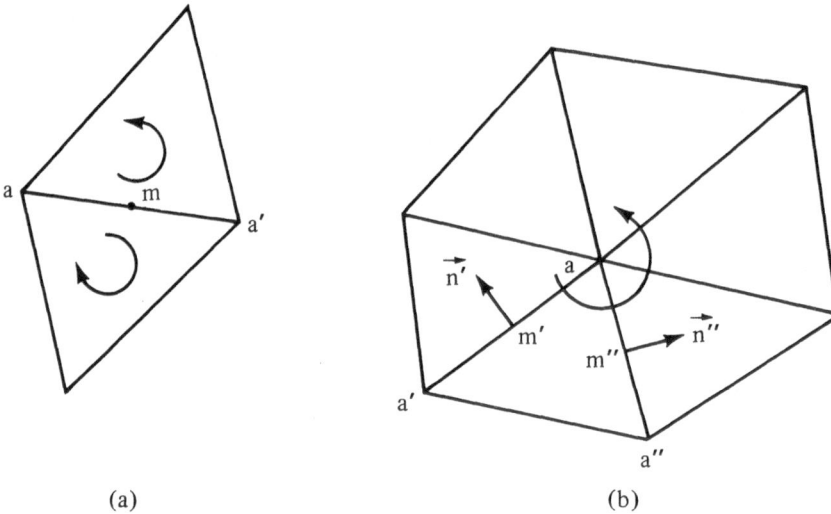

(a)                                                          (b)

Fig. 28

*Case 2:* no obstacle in the flow, non homogeneous boundary condition (e.g.: given profile of velocity at the entrance of a channel), then the $U_a$ are determined up to an additive constant, so that we can pick a vertex point $a_0$ and fix $U_{a_0}=0$.

*Case 3:* one or several *obstacles* in the flow (e.g. an airfoil); we assume: $\mathbf{u}=0$ on the obstacles. Then all the $U_a$ corresponding to vertices on one obstacle are equal. Thus we see another type of (non local) basis functions associated to each obstacle $\Gamma_i$:

$$\mathbf{w}_{\Gamma_i} = \sum_{a \in \Gamma_i} \mathbf{w}_a$$

(This represents a circulation around the obstacle).

Now we insert (108) into the discrete analogue of (99):

$$\sum_T \int_T \nu\,\mathbf{grad}\,\mathbf{u}_h \cdot \mathbf{grad}\,\mathbf{v}_h\,dx = \int_\Omega \mathbf{f}\mathbf{v}_h\,dx$$

with $\mathbf{v}_h = \mathbf{w}_a, \mathbf{w}_m$ for all interior nodes (and $\mathbf{w}_{\Gamma_i}$ in case 3).

Note that the number of equations (=number of degrees of freedom) is about the same as for the approximation of Laplace's equation by *P*2 triangles, i.e.: number of vertices + number of mid side nodes.

**Remarks.**

(i) The properties of the coefficients $U_a$ can be enlightened if we note that $U_a$ is the value, at point $a$, of the approximate streamfunction associated to $\mathbf{u}_h$;

(ii) For further details on the implementation see Thomasset (1976, 1980);

(iii) The set $\{w_s, w_m, w_{\Gamma_i}\}$ was proved to be a base in Thomasset (1977); in Appendix 3, we give a more elegant proof due to F. Hecht; this proof can be extended to construct a divergence basis in 3 dimensions (Hecht (1980)).

### 3.2.2. $u_h$: $P2$ Triangle[56] $p_h$: $P0$ (=Piecewise Constant)

The degrees of freedom are: for the velocity, the values at vertices and mid side nodes (the velocity is continuous in this approximation); for the pressure, any point in each triangle $T$, e.g. the barycentre $O^T$. Note that the discretization of the pressure gradient term yields from (105) (since $\operatorname{div} v_h$ is a polynomial of degree $\leq 1$):

$$\sum_T \int_T p_h \operatorname{div} v_h \, dx = \sum_T \operatorname{area}(T) p_h(O^T)(\operatorname{div} v_h(O^T)).$$

So that the divergence of $v_h$ has the same degrees of freedom as $p_h$; the incompressibility condition is to satisfied in the mean on each triangle:

$$\operatorname{div} u_h(O^T) \equiv \frac{3}{\operatorname{area}(T)} \int_T \operatorname{div} u_h \, dx = 0$$

*Error estimates:* (Fortin (1972), Crouzeix and Raviart (1973)).

$$|u - u_h|_{1,h} = O(h), \quad |p - p_h|_0 = O(h)$$

This result is disappointing since the same order of accuracy can be obtained with the $P1$-non-conforming triangle, the implementation of which is much simpler (however the situation is reversed in axisymmetric coordinates: the "$P2 - P0$" f.e.m should be preferred in this case (Lailly (1976), Legait (1980)), the implementation being easier).

The above error estimate is the motivation for the following finite element.

### 3.2.3. $u_h$: "$P2$+bubble" Triangle (or Modified $P2$); $p_h$: Discontinuous $P1$

We enrich the polynomials of degree 2 with one polynomial of degree 3 in order that the Brezzi-Babuška's condition can be satisfied with pressures of degree $\leq 1$. Namely, for each triangle $T$, we let $\lambda_1, \lambda_2, \lambda_3$ be the barycentric coordinates and:

$$\beta^T(x) = 27\lambda_1(x)\lambda_2(x)\lambda_3(x).$$

The "bubble" function $\beta^T(x)$ is devised to vanish along the triangle sides and normalized to 1 at the barycentre.

Each component of $u_h$ is required to be of the form, on a triangle $T$:

$$u_{hi} = (\text{any polynomial of degree } 2) + \alpha_i^T \beta^T(x)$$

---

[56] I.e. the 6 node triangular lagrangian element, see §1-4-2.

Since in addition $u_h$ is required to be continuous at the interelement boundaries, the degrees of freedom for component $u_{hi}$ are:

The values of $u_{hi}$ at the vertices and mid side nodes;

On each triangle, a coefficient $\alpha_i^T$ (which differs from the value at the barycentre because the shape functions associated to vertices or mid side nodes do not vanish at the barycentre), of the bubble function.

The discrete pressure is piecewise polynomial of degree $\leq 1$; since no requirements are made as to the continuity of $p_h$, the degrees of freedom of $p_h$ can be the values at three arbitrary points on the triangle.

*Error estimates:* Crouzeix and Raviart (1973)

$$|u-u_h|_{1,h} = O(h^2)$$

$$|p-p_h|_0 = O(h^2)$$

Thus the best possible accuracy (using modified $P2$ triangles) is obtained.

### 3.2.4. $u_h$: $Q2$ Quadrangle[57]; $p_h$: $Q1$ Discontinuous

Bercovier (1978), Bercovier and Engelman (1979), Engelman (1979).

This is the quadrangle analogue of the preceding method; with respect to the curvilinear coordinates $\xi_1, \xi_2$ (the coordinates in the reference element), the velocity is biquadratic and the pressure (discontinuous) bilinear. The degrees of freedom are: for the velocity the values at the vertices, "midside" nodes, and centre nodes; for the pressure, the values at 4 points within the elements.

*Error estimates:*

$$|u-u_h|_{1,h} = O(h^2)$$

**Remark.**

(i) The Brezzi' condition fails to be true for this element (P. Le Tallec, personal communication); therefore computed pressures may suffer oscillations ("checkerboard pattern").[58]

(ii) (Raviart (1979)): equation (94) is equivalent to (95) only when the divergence of **u** is exactly 0; in all the methods of this section we worked with velocity fields for which an approximation of the divergence is 0. Therefore it should be better in practice to use instead of (96), the following formulation

---

[57] I.e. the 9 node Lagrange quadrilateral element, see §1-4-4.

[58] Cf. Hughes, Liu and Brooks (1979) such oscillations may be corrected by smoothing procedures.

directly derived from (94):

Find $\mathbf{u} \in \left( H_0^1(\Omega) \right)^2$ and $p \in L^2(\Omega)$ such that:

$$\int_\Omega \frac{\nu}{2} (u_{i,j} + u_{j,i})(v_{i,j} + v_{j,i}) \, dx + \int_\Omega p \operatorname{div} \mathbf{v} \, dx$$

$$= \int_\Omega \mathbf{f} \cdot \mathbf{v} \, dx, \qquad \forall \, v \in \left( H_0^1(\Omega) \right)^2 \tag{109}$$

$$\int_\Omega q \operatorname{div} \mathbf{u} \, dx = 0, \qquad \forall \, q \in L^2(\Omega)$$

However we shall continue, for simplicity, to use the formulation derived from (95) such as (96), since the modifications necessary to transpose the methods to (109), are of small importance. Formulation (109) is used e.g. by Hughes et al. (1979); nevertheless the precise effect on the accuracy of the use of (109) instead of (96), is not known to the author at the moment of this writing.

### 3.2.5. Numerical Solution by Penalty Methods

Bercovier (1978), Fortin and Glowinski (to appear), Hughes, Levy and Taylor (1976), Bercovier and Engelman (1975), Engelman (1979).

Although penalty[59] methods may be used for general saddle point problems (Bercovier (1974)), one of their most useful applications can be found in the treatment of the incompressibility condition, with a piecewise discontinuous approximation of the pressure. The principle is to replace the constraint $\operatorname{div} \mathbf{u} = 0$, by a relation between $\mathbf{u}$ and $p$ of the form: $+\varepsilon p + \operatorname{div} \mathbf{u} = 0$, where $\varepsilon$ is a small parameter.

More precisely we consider the following discrete equations ($M_h$ and $X_h$ are the spaces of approximation for $\mathbf{u}$ and $p$ encountered above):

a)    Find $\mathbf{u}_{h,\varepsilon} \in M_h, p_{h,\varepsilon} \in X_h$ such that

$$\nu \int_\Omega \operatorname{grad} \mathbf{u}_{h,\varepsilon} \cdot \operatorname{grad} \mathbf{v}_h \, dx - \int_\Omega p_{h,\varepsilon} \operatorname{div} \mathbf{v}_h \, dx$$

$$= \int_\Omega \mathbf{f} \cdot \mathbf{v}_h \, dx, \qquad \forall \, \mathbf{v}_h \in M_h \tag{110}$$

b)    $\varepsilon \int_\Omega p_{h,\varepsilon} q_h \, dx + \int_\Omega q_h \operatorname{div} \mathbf{u}_{h,\varepsilon} \, dx = 0, \qquad \forall \, q_h \in M_h$

---

[59] The term of penalty is to be understood in the framework of optimal control: the cost functional is augmented with $\varepsilon^{-1} \int |\operatorname{div} v|^2 \, dx$, so that diverging velocity fields are strongly penalized in the minimization process.

(e.g., $M_h$ = set of continuous piecewise $P2$ vector functions vanishing on the boundary, $X_h$ = set of discontinuous piecewise constant functions, as in §3-2-2).

The following orders of accuracy can be proved (Bercovier (1978))

$$|u - u_{h,\varepsilon}|_{1,h} + |p - p_h|_0 = O(\varepsilon) + O(h^l) \tag{111}$$

where $O(h^l)$ is the same order of accuracy as in the method without penalty ($\varepsilon = 0$), that is:

$l = 1$   for   $\mathbf{u}_h$: $P1$ non conforming or $P2$

$\qquad\qquad p_h$: $P0$ discontinuous

$l = 2$   for   $\mathbf{u}_h$: "$P2$+bulb", $p_h$: $P1$ discontinuous

$\qquad\qquad \mathbf{u}_h$: $Q2$, $p_h$: $Q1$ discontinuous

Now the practical interest of (110) appears if we can eliminate easily the pressure $p_{h,\varepsilon}$ between (110a) and (110b), so that the *number of equations will be reduced*. Since the discrete pressure is discontinuous this elimination can be done locally, triangle per triangle.

EXAMPLE 1. $\mathbf{u}_h$: $P2$, $p_h$: discontinuous $P0$
In this case, (110b) reduces to:

$$p_{h,\varepsilon}(T) = -\frac{1}{\varepsilon}(\operatorname{div}\mathbf{u}_{h,\varepsilon})(O^T) \tag{112}$$

for each triangle $T$ ($O^T$ = barycentre of $T$).

Reporting (112) to (110a) we find the discrete equations:

$$\nu \int_\Omega \operatorname{grad}\mathbf{u}_{h,\varepsilon} \cdot \operatorname{grad}\mathbf{v}_h \, dx$$
$$+ \frac{1}{\varepsilon} \sum_T \operatorname{area}(T)(\operatorname{div}\mathbf{u}_{h,\varepsilon})(O^T)(\operatorname{div}\mathbf{v}_h)(O^T) \tag{113}$$
$$= \int_\Omega \mathbf{f} \cdot \mathbf{v}_h \, dx, \quad \forall \, \mathbf{v}_h \in M_h$$

The situation is the same when $P1$ non conforming elements are used for $\mathbf{u}_h$.

EXAMPLE 2. $u_h$: "$P2$+bubble", $p_h$: discontinuous $P1$
The integral $\int_T p_h q_h \, dx$ is exactly integrated by the 3-point formula with integration points at the mid side nodes; therefore if we choose these points as nodal points for the pressure the dependence of $p_{h,\varepsilon}$ on $\operatorname{div}\mathbf{u}_{h,\varepsilon}$ in (110b) is explicit:

$$p_{h,\varepsilon}|_T = \sum_{m \in T} p_{h,\varepsilon}^m \phi_m \quad \text{(where } \phi_m \text{ is the shape function associated to mid side node } m)$$

Then:

$$p_{h,\varepsilon}^m = \frac{1}{3} \times \frac{3}{\text{area}(T)} \int_T \phi_m \operatorname{div} \mathbf{u}_{h,\varepsilon} \, dx,$$

and we can put these values into (110a); the equations obtained will be slightly more complicated than (113).

EXAMPLE 3. $\mathbf{u}_h$: $Q2$, $p_h$: discontinuous $Q1$.

The situation is indeed very much the same as in the previous example. We choose, as degrees of freedom of the pressure on each element, the values at the form Gauss integration points, i.e. with curvilinear coordinates:

$$\frac{1 \pm 1/\sqrt{3}}{2}, \frac{1 \pm 1/\sqrt{3}}{2}$$

$(g_m^K = F_K(\hat{g}_m))$.

Then when $F_K \in Q_1 \times Q_1$ [60]:

$$\int_K p_h q_h \, dx = \int_{\hat{K}} J_K (p_h \circ F_K)(q_h \circ F_K) \, d\xi$$

$$= \sum_{m=1}^4 J_K(\hat{g}_m) p_h(g_m) q_h(g_m)$$

The above expression further simplifies in the case of a *rectangular* element: then the jacobian $J_K$ is constant and $(p_h \operatorname{div} \mathbf{u}_h)$ is a polynomial of degree at most 3; which is exactly by the $2 \times 2$ Gauss rule. Therefore (110b) yields in this case:

$$p_{h,\varepsilon}(g_m^K) = -\frac{1}{\varepsilon} \operatorname{div} \mathbf{u}_{h,\varepsilon}(g_m^K), \qquad m = 1,2,3,4$$

and the discrete equations become:

$$\nu \int_\Omega \operatorname{grad} \mathbf{u}_{h,\varepsilon} \cdot \operatorname{grad} \mathbf{v}_h \, dx + \frac{1}{\varepsilon} \sum_K J_K \sum_{m=1}^4 \operatorname{div} \mathbf{u}_{h,\varepsilon}(g_m^K) \operatorname{div} v_h(g_m^K)$$

$$= \int_\Omega \mathbf{f} \mathbf{v}_h \, dx, \qquad \forall, \ \mathbf{v}_h \in M_h$$

**Remarks.**

(i) *Selection of parameter $\varepsilon$:* from the error estimate (111), $\varepsilon$ should be small enough so that the compressibility errors are small compared to the interpolation errors; yet $\varepsilon$ should not be so small that numerical ill conditioning ensues. From the experiments of Engelman (1979) it appears that a value of $\varepsilon = 10^{-2} h^2$ (for the "$Q1 - Q1$" element), $l = 2$) yields an appropriate accuracy.

---

[60] I.e. for an element with straight sides.

Hughes *et al.* (1979) pointed out that the choice of $\varepsilon$ should be made
dependent upon the word length of the computer.

(ii) A general saddle point problem can be put into the matrix form:

$$\begin{bmatrix} A & B^T \\ B & 0 \end{bmatrix}\begin{bmatrix} U \\ P \end{bmatrix}=\begin{bmatrix} F \\ G \end{bmatrix}$$

Then the penalty method applied to this system, will result into the following
equations:

$$\begin{bmatrix} A & B^T \\ B & -\varepsilon C \end{bmatrix}\begin{bmatrix} U \\ P \end{bmatrix}=\begin{bmatrix} F \\ G \end{bmatrix}$$

where $C$ is some positive definite matrix; elimination of $P$ yields:

$$(A+\varepsilon^{-1}B^TC^{-1}B)U=F-\varepsilon^{-1}B^TC^{-1}G$$

This is practically useful only if $C$ is a diagonal (or block-diagonal) matrix as
in the methods of this §3-2.

(iii) When the velocity has been computed, the *pressure* can be retrieved
through: $p_{h,\varepsilon}=-1/\varepsilon\,\mathrm{div}\,\mathbf{u}_{h,\varepsilon}$, or some equivalent equation; this discontinuous approximation of the pressure suffers from oscillations; it can be im-
proved through smoothing procedures, as described by Hughes *et al.* (1979).

(iv) *Duality methods:* as we have noticed, Stokes equations are equivalent
to a saddle point problem (99)

$$\underset{\mathbf{v}}{\mathrm{Min}}\ \underset{q}{\mathrm{Max}}\ \mathcal{L}(\mathbf{v},q)$$

So that a possible algorithm for the numerical solution is as follows:

Uzawa's Algorithm:

$p^0$ arbitrary initial guess.

For $m\geq0$:

Find $u^m$ such that

$$\mathcal{L}(\mathrm{u}^m,\mathrm{p}^m)=\underset{v}{\mathrm{Min}}\ \mathcal{L}(\mathrm{v},\mathrm{p}^m)$$

$$p^{m+1}=p^m+\rho\frac{\partial\mathcal{L}}{\partial p}(\mathrm{u}^m,\mathrm{p}^m)$$

In the case of Stokes equations, the equation in $u^m$ reduces to 2 Poisson's
equations. The pressure equation yields, when for instance piecewise $P0$
approximation is used for the pressure

$$p_h^{m+1}=p_h^m-\rho\,\mathrm{div}\,u_h^{m+1}.^{[61]}$$

---

[61] Note that the pure penalty algorithm coincides with the first Uzawa iteration with $p^0=0$
and $\rho=1/\varepsilon$.

The algorithm converges for $0<\rho<2\nu/d$ ($d=$dimension of space$=2$ or 3). The convergence rate is improved by several orders of magnitude, with the addition of a *penalty* term to the Lagrangian:

$$\mathcal{L}_\varepsilon(\mathbf{v}, q) = \mathcal{L}(\mathbf{v}, q) + \frac{1}{2\varepsilon}|\operatorname{div}\mathbf{v}|^2$$

$$= \frac{1}{2}\int_\Omega \left(\nu|\operatorname{grad}v|^2 + \frac{1}{\varepsilon}|\operatorname{div}\mathbf{v}|^2 - 2f\mathbf{v}\right)dx$$

$$- \int_\Omega p\operatorname{div}\mathbf{v}\,dx.$$

The application of Uzawa's algorithm yields at each step, an equation of the form:

$$-\nu\Delta u^{m+1} - 1/\varepsilon\operatorname{grad}\operatorname{div}u^{m+1} + \nabla p^m = f$$

The parameter $\rho$ can be chosen in a larger interval: $0<\rho<2(1/\varepsilon+\nu/d)$; the choice of $\rho$ can be done according to the simple rule: $\rho=1/\varepsilon$ (which is nearly optimal).

*Ref.*: Fortin and Glowinski (1980), Glowinski and Pironneau (1979), Temam (1977), Thomasset (1976).

### 3.2.6. Numerical Results and Further Comments

The above theoretical results on the accuracy are supported by the experiments of Segal[62] (1978) for the triangular elements; of Engelman (1979) for quadrangular ($Q2-Q2$) elements. The last reference shows the dramatic improvement brought in by the use of complete $Q2$ elements for **u**, instead of the 8-node incomplete $Q2$ (serendip) element used by Zienckiewicz and Godbole (1975; who also used a penalty formulation).[63]

We note that all the methods of this section can be extended to three dimensional problems; the chief difficulties arise in the 3-D extension of the divergence free basis (F. Hecht (1980)). The "$P2-P0$" element is used in axisymmetric coordinates by Legait (1980). Other numerical results will be discussed in chapter 4.

---

[62] From which it appears that the $P2+P0$ element is slightly better than the "(non conforming) $P1+P0$".

[63] A variant of Uzawa's algorithm for Navier Stokes equations was early studied by Chorin (1967, 1968) and recently by Felippa (1978). Other penalty formulations are considered by Zienckiewcz (1977), Boisvert (1979).

## 3.3. Velocity–Pressure Formulations: Continuous Approximation of the Pressure and Velocity

### 3.3.1. Introduction

The weak formulation is readily obtained from (95):

Find $\mathbf{u}$   and   $p$ such that                                            (114)

a)   $\nu \int_\Omega \mathbf{grad}\,\mathbf{u} \cdot \mathbf{grad}\,\mathbf{v}\,dx + \int_\Omega \mathbf{grad}\,p \cdot \mathbf{v}\,dx$

$= \int_\Omega \mathbf{f} \cdot \mathbf{v}\,dx, \quad \forall\, \mathbf{v} \in H_0^1(\Omega) \times H_0^1(\Omega)$

b)   $\int_\Omega q\,\mathrm{div}\,\mathbf{u}\,dx = 0, \quad \forall\, q \in H^1(\Omega)$

In the approximations derived from (114) the velocity *flux* across any element is small but a priori non vanishing.

We note that the structure of the matrix of finite element equations obtained from (114) is of the form:

$$\begin{bmatrix} A & B^T \\ B & 0 \end{bmatrix}$$

and that the penalization technique is inoperant in this case: $B$ has a non local structure. Thus Taylor and Hood (1973) required the use of a *frontal method*[64] to solve their linear systems. The *chief interest* of such a formulation as (114), is to allow for a *decomposition* of the problem into a succession of Poisson's equations, thus leading to an efficient solver of Stokes (and consequently Navier Stokes) equations. This technique is reviewed in §3-3-3.

Next section describes some examples and the error estimates obtained by Bercovier and Pironneau (1979).

### 3.3.2. Examples and Error Estimates

(Bercovier and Pironneau (1979)).

The formulation (114) contains derivatives of the pressure; therefore we require *conforming* elements for the approximation of $p$ (i.e., $p_h$ is continuous all over the domain).

EXAMPLE 1. $\mathbf{u}_h$: $P_2$-triangle (6-nodes-triangle)

$p_h = P_1$-triangle (3-nodes-triangle)

---

[64] A special version of Gauss elimination for solving finite element equations, using out of core storage: Irons (1970), Hood (1976).

*Error estimates*[65]:

$$|u-u_h|_1 = \left( \int_\Omega |\text{grad}(u-u_h)|^2 \, dx \right)^{1/2} = O(h^2)$$

$$|p-p_h|_1 = \left( \int_\Omega |\text{grad}(p-p_h)|^2 \, dx \right)^{1/2} = O(h)$$

EXAMPLE 2. $\mathbf{u}_h = Q_2$-quadrangle (9-nodes-element)

$p_h$: $Q_1$-quadrangle (4-nodes-element)

*Error estimates:* same as in example 1.

EXAMPLE 3. $\mathbf{u}_h$: $4 \times P_1$ triangle

$p_h$: $P_1$-triangle

The meaning of the notation "$4 \times P_1$" triangle for the velocities' approximation is as follows: if $\mathcal{T}_h$ is the triangulation used for the approximation of the pressure, we let $\mathcal{T}_{h/2}$ be the triangulation obtained by subdividing each triangle of $\mathcal{T}_h$ into 4 sub-triangles, and we require $\mathbf{u}_h$ to be polynomial of degree $\leq 1$ on $\mathcal{T}_{h/2}$. We note that the number of degrees of freedom is the same as in example 1 ($P2$-triangle for $\mathbf{u}_h$ on the triangulation $\mathcal{T}_h$, $P1$-triangle for $p_h$ on the same triangulation).

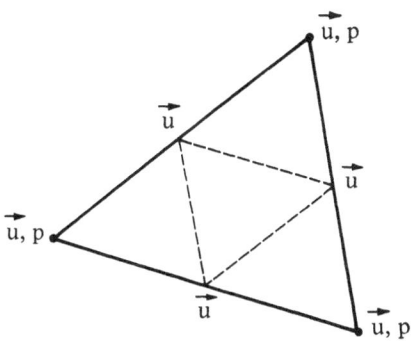

*Error estimates:* $|u-u_h|_1 = O(h)$

$$|p-p_h|_0 = \left( \int_\Omega |p-p_h|^2 \, dx \right)^{1/2} = O(h)$$

Thus the implementation (the numerical integrations) is simpler than in example 1, but one order of accuracy is lost.

---

[65] The assumption is made that every triangle in the mesh has at most two vertices belonging to the boundary.

EXAMPLE 4. $\mathbf{u}_h$: $4 \times Q_1$ quadrangle

$p_h$: $Q_1$ quadrangle

This is, with obvious notations, the quadrangle analogue of example 2.

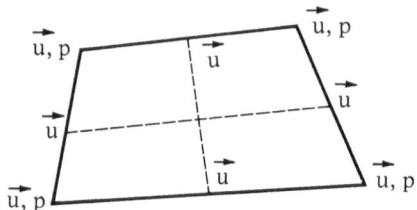

*Error estimates:* same as in example 2.

### 3.3.3. Decomposition of the Stokes Problem

(Glowinski and Pironneau (1979)).

The pressure and velocity fields are approximated as in the preceding section (i.e.: $\mathbf{u}_h$: $P2$, $p_h = P1$, etc). The method proposed by R. Glowinski and O. Pironneau produces a decomposition of the Stokes problem into a cascade of Poisson's equations and yields efficient algorithms for the numerical solution of (2D or 3D) Stokes (and Navier Stokes, see Chapter 4) equations.

For a better understanding of the method we begin with the continuum equations:

a)    $-\nu \Delta \mathbf{u} = \mathbf{f} - \mathbf{grad}\, p$

b)    $\operatorname{div} \mathbf{u} = 0$                                                    (115)

c)    $\mathbf{u}|_\Gamma = 0$

Assume that we happen to know the values of the pressure at all points on the boundary; we note $\lambda$ this boundary-defined function. Then the whole pressure field can be obtained through the solution of a Poisson's equation, which is formed by applying the operator div to (115a):

$$\begin{cases} -\Delta p = -\operatorname{div} f \\ p|_\Gamma = \lambda \end{cases} \tag{116}$$

Then we get the velocities by the solution of another set of Poisson's equations:

$$\begin{cases} -\nu \Delta \mathbf{u} = \mathbf{f} - \mathbf{grad}\, p \\ \mathbf{u}|_\Gamma = 0 \end{cases} \tag{117}$$

Of course the exact value of $\lambda$, the boundary pressure is not known a priori; and if we start from an arbitrary $\lambda$, there is no reason for the velocity field obtained through (116), (117) to be divergence free. We know that this velocity field admits decomposition of the form:

$$\mathbf{u} = \operatorname{grad} \phi + \operatorname{curl} \psi$$

($\psi$: a vector in 3D).

We can choose the potential $\phi$ to be zero on the boundary; then this $\phi$ is obtained from $\mathbf{u}$ by another Poisson's equation:

$$\begin{cases} -\Delta\phi = -\operatorname{div}\mathbf{u} \\ \phi|_\Gamma = 0 \end{cases} \tag{118}$$

From (116), (117), (118) we see that:

$$\Delta\Delta\phi = \Delta(\operatorname{div}\mathbf{u}) = \operatorname{div}(\Delta\mathbf{u}) = \nu^{-1}\operatorname{div}(f - \operatorname{grad} p) = 0$$

We must look at $\phi$ as a function of $\lambda$: if we *can* find $\lambda$ such that $\partial\phi/\partial n|_\Gamma = 0$, then $\phi$ satisfies a biharmonic equation with homogeneous data:

$$\begin{cases} \Delta\Delta\phi = 0 \\ \phi|_\Gamma = 0 \\ \dfrac{\partial\phi}{\partial n} = 0 \end{cases}$$

which implies $\phi = 0$, and therefore $\operatorname{div}\mathbf{u} = 0$. Thus our conclusion is that $\partial\phi/\partial n|_\Gamma = 0$ is a necessary and sufficient condition for the incompressibility constraint to be satisfied. If we set:

$$\mathcal{Q}\lambda = \frac{\partial\phi}{\partial n} \tag{119}$$

we have to solve a linear affine equation for boundary valued functions $\lambda$:

$$\mathcal{Q}\lambda = 0 \tag{120}$$

Let us *rewrite* these equations by introducing the solutions of the following Poisson's equations:

$$\begin{cases} -\Delta p_0 = -\operatorname{div} f \\ p_0|_\Gamma = 0 \end{cases} \tag{121}$$

$$\begin{cases} -\nu\Delta\mathbf{u}_0 = f - \operatorname{grad} p_0 \\ \mathbf{u}_0|_\Gamma = \mathbf{0} \end{cases} \tag{122}$$

$$\begin{cases} -\Delta\phi_0 = -\operatorname{div}\mathbf{u}_0 \\ \phi_0|_\Gamma = 0 \end{cases} \tag{123}$$

Set $\tilde{p} = p - p_0$, $\tilde{\mathbf{u}} = \mathbf{u} - \mathbf{u}_0$, $\tilde{\phi} = \phi - \phi_0$; $\tilde{\lambda} = \tilde{p}|_\Gamma$. Then we have:

$$\begin{cases} -\Delta\tilde{p} = 0 \\ \tilde{p}|_\Gamma = \tilde{\lambda} \end{cases} \tag{124}$$

$$\begin{cases} -\nu\Delta\tilde{\mathbf{u}} = -\operatorname{grad}\tilde{p} \\ \tilde{u}|_\Gamma = 0 \end{cases} \tag{125}$$

$$\begin{cases} \Delta\tilde{\phi} = \operatorname{div}\tilde{u} \\ \tilde{\phi}|_\Gamma = 0 \end{cases} \tag{126}$$

(so that $\tilde{p}, \tilde{\mathbf{u}}, \tilde{\phi}$ are function of $\tilde{\lambda}$).

Set:      $\mathcal{B}\tilde{\lambda} \equiv + \dfrac{\partial \tilde{\phi}}{\partial n}$

The condition $\operatorname{div} \mathbf{u} = 0$ is equivalent to:

$$\frac{\partial}{\partial n}(\tilde{\phi} + \phi_0) = 0$$

that is

$$\boxed{\mathcal{B}\tilde{\lambda} = -\frac{\partial \phi_0}{\partial n}} \tag{127}$$

The above mentioned authors proved that $\tilde{\lambda} \to \mathcal{B}\tilde{\lambda}$ is an invertible one-to-one (provided $\int_\Gamma \tilde{\lambda}\, ds = 0$) operator so that (127) has a unique solution. Moreover the bilinear form associated to $\mathcal{B}$:

$$b(\tilde{\lambda}, \mu) = \langle \mathcal{B}\tilde{\lambda}, \mu \rangle \; ^{66}$$

is symmetric and positive definite.

We note that the right hand side in (127) can be transformed as follows, by the Green's formula:

$$\int_\Gamma \frac{\partial \phi_0}{\partial n} \mu\, ds = \int_\Omega \nabla \phi_0 \nabla \bar{\mu}\, dx + \int_\Omega \Delta \phi_0 \bar{\mu}\, dx.$$

(if $\bar{\mu}$ is defined over $\Omega$ and such that $\bar{\mu}|_\Gamma = \mu$)

$$\int_\Gamma \frac{\partial \phi_0}{\partial n} \mu\, ds = \int_\Omega (\nabla \phi_0 - \mathbf{u}_0) \cdot \nabla \bar{\mu}\, dx \tag{128}$$

(from (123) and another application of the Green's formula).

Now we come to the *finite element implementation* of this method; for the simplicity of our description, we assume that we use the "$P2 - P1$" interpolation; the extension to other types of elements is straightforward. We require some notations (for a given triangulation $\mathcal{T}_h$):

$V_{h0}$ = approximation space of velocities
      = set of continuous, piecewise polynomial of degree $\leq 2$, vector valued functions, vanishing on $\Gamma$.

$W_h$ = approximation space of pressures
      = set of continuous, piecewise polynomial of degree $\leq 1$

$W_{h0}$ = set of functions in $W_h$, vanishing on $\Gamma$.

We introduce the basis functions of $W_h$: $\phi_i$ is the basis function associated to vertex $i$; thus $W_h$ is generated by the basis functions associated to *interior* vertices.

---

[66] $\langle \cdot, \cdot \rangle$: duality pairing between $H^{1/2}(\Gamma)$ and $H^{-1/2}(\Gamma)$ (the integral over $\Gamma$ for sufficiently smooth $\tilde{\lambda}$ and $\mu$).

We set $\mathfrak{M}_h$: subspace of $W_h$ generated by those basis functions associated to vertices on the *boundary*. Thus $W_h$ is the direct sum of $W_{h0}$ and $\mathfrak{M}_h$:

$$W_h = W_{h0} \oplus \mathfrak{M}_h$$

($\mathfrak{M}_h$ will be our approximation space for the boundary pressure $\lambda$).

Define the discrete analogues of Poisson's equations (121) to (123).

$$\begin{cases} p_{h0} \in W_{h0} \\ \displaystyle\int_\Omega \mathbf{grad}\, p_{h0} \cdot \mathbf{grad}\, w\, dx = -\int_\Omega w(\mathrm{div}\,\mathbf{f})\, dx, & \forall\, w \in W_{h0} \end{cases} \tag{129}$$

$$\begin{cases} \mathbf{u}_{h0} \in V_{h0} \\ \displaystyle\int_\Omega \mathbf{grad}\, \mathbf{u}_{h0} \cdot \mathbf{grad}\, \mathbf{w}\, dx = \int_\Omega (\mathbf{f} - \mathbf{grad}\, p_{h0}) \cdot \mathbf{w}, & \forall\, \mathbf{w} \in V_{h0} \end{cases} \tag{130}$$

$$\begin{cases} \phi_{h0} \in W_{h0} \\ \displaystyle\int_\Omega \mathbf{grad}\, \phi_{h0} \cdot \mathbf{grad}\, w\, dx = -\int_\Omega w\,\mathrm{div}\,\mathbf{u}_0\, dx, & \forall\, w \in W_{h0} \end{cases} \tag{131}$$

We now have to construct the discrete analogue of operator $\mathfrak{B}$: this is an $M \times M$ matrix ($M$ being the number of vertices on $\Gamma$, i.e. the dimension of $\mathfrak{M}_h$). For any $\hat{\lambda}_h$ in $\mathfrak{M}_h$ we define the discrete analogues of (124), (125), (126):

$$\tilde{p}_h - \tilde{\lambda}_h \in W_{h0}$$

$$\int_\Omega \mathbf{grad}\, \tilde{p}_h \cdot \mathbf{grad}\, w\, dx = 0, \qquad \forall\, w \in W_{h0} \tag{132}$$

$$\tilde{\mathbf{u}}_h \in V_{h0}$$

$$\nu \int_\Omega \mathbf{grad}\, \tilde{\mathbf{u}}_h \cdot \mathbf{grad}\, \mathbf{w}\, dx = -\int_\Omega \mathbf{grad}\, \tilde{p}_h \cdot \mathbf{w}\, dx, \qquad \forall\, \mathbf{w} \in V_{h0} \tag{133}$$

$$\tilde{\phi}_h \in W_{h0}$$

$$\int_\Omega \mathbf{grad}\, \tilde{\phi}_h \cdot \mathbf{grad}\, w\, dx = -\int_\Omega w\,\mathrm{div}\,\tilde{\mathbf{u}}\, dx, \qquad \forall\, w \in W_{h0} \tag{134}$$

Thus $\tilde{\phi}_h = \tilde{\phi}_h(\tilde{\lambda}_h)$; we would like to find a $\tilde{\lambda}_h$ such that, in some sense, the normal derivative of $(\tilde{\phi}_h + \phi_{h0})$ vanishes:

$$\frac{\text{``}\partial\text{''}}{\partial n}(\tilde{\phi}_h + \phi_{h0}) = 0.$$

Although pointwise values of normal derivatives for $\tilde{\phi}_h$ and $\phi_{h0}$ can be computed they cannot be considered as realistic approximations to normal derivatives of the true solution. So that a weak form of this condition should be used. In view of (128), we impose instead the following condition:

$$\int_\Omega \left( \mathbf{grad}\, \tilde{\phi}_h \cdot \mathbf{grad}\, \mu_h - \mu_h \operatorname{div} \tilde{\mathbf{u}}_h \right) dx$$

$$= - \int_\Omega \left( \mathbf{grad}\, \phi_{h0} \cdot \mathbf{grad}\, \mu_h - \mu_h \operatorname{div} \mathbf{u}_{h0} \right) dx, \qquad \forall\, \mu_h \in \mathfrak{M}_h \qquad (135)$$

This set of equations will determine $\tilde{\lambda}_h$: it is the discrete analogue of (127). Now we know how to construct the matrix analogue of operator $\mathfrak{B}$: in order to compute the column $i$, associated to boundary vertex $i$ and basis function $\phi_i \in \mathfrak{M}_h$, we put $\tilde{\lambda}_h = \phi_i$ in (132) and successively compute the corresponding $\tilde{p}_h, \tilde{\phi}_h, \tilde{\mathbf{u}}_h$; the coefficient in column $i$ and line $j$ is obtained by putting $\mu_h = \phi_j$ in (135):

$$b_{ij} = \int_\Omega \left( \mathbf{grad}\, \tilde{\phi}_h(\phi_i) \cdot \mathbf{grad}\, \phi_j - \phi_j \operatorname{div} \tilde{\mathbf{u}}_h(\phi_i) \right) dx$$

(so that, to compute the $M$ [67] coefficients of only *one* column of matrix $[b_{ij}]$ we have to solve: 4 Poisson's equations for a 2-dimensional problem; 5 Poisson's equations for a 3-dimensional problem). Once this matrix is formed we have to solve equations (135), that is:

$$\tilde{\lambda}_h = \sum_{i=1}^M \Lambda_i \phi_i$$

$$\sum_{i=1}^M \Lambda_i b_{ij} = \int_\Omega \left( \phi_j \operatorname{div} \mathbf{u}_{h0} - \mathbf{grad}\, \phi_{h0} \cdot \mathbf{grad}\, \phi_j \right) dx, \qquad j = 1, \ldots, M$$

(Note that the support of $\phi_j$ is close to the boundary). The matrix $[b_{ij}]$ is proved to be:

Symmetric
Positive definite.

Therefore the above system can be solved either by Cholesky decomposition or by conjugate gradient method; unfortunately this system is not sparse.

Once this relatively small system is solved, yielding the solution $\tilde{\lambda}_h$, the complete solution is found via the further solution of 4 (5 in 3D) Poisson's equations, and:

$$\phi_h = \tilde{\phi}_h + \phi_{0h}, \qquad \mathbf{u}_h = \tilde{\mathbf{u}}_h + \mathbf{u}_{0h}, \qquad p_h = \tilde{p}_h + p_{0h}$$

---

[67] Remind that $M$ is the number of boundary vertex nodes.

*Error estimates:* (piecewise quadratic velocity, piecewise linear pressure):

$$|u_h - u|_1 = O(h^2)$$
$$|p_h - p|_1 = O(h)$$
$$|\phi_h|_1 = O(h^2)$$

(the $H_1$ norm of $\phi_h$ measures the $L2$ norm of the divergence div $u_h$).

**Remark.** If we chose to impose the incompressibility through $\phi_h \equiv 0$ instead of (135) we find exactly the methods of §3-3-2 employed e.g. by Taylor and Hood (1973). (Note that $\phi_h = 0$ does not imply the vanishing of div $u_h$, contrary to the continuous case).

The present method reduces the Stokes problem to the solution of a small finite number of discrete Poisson's equations, plus the solution of a discrete boundary integral equation whose matrix is symmetric and positive definite. This method appeared to be an efficient Stokes solver when applied to the iterative solution of Navier Stokes equations, requiring repeated solution of Stokes problems (see Chapter 4 of this course; let us insist now on the fact that in such a process the matrix $[b_{ij}]$ needs to be computed and factorized only once).

Finally we note that the method was originated in decomposition problems for the biharmonic equation: see Glowinski and Pironneau (1977) (the method is briefly reviewed in §3-5).

## 3.4. Vorticity–Pressure–Velocity Formulations: Discontinuous Approximations of Pressure and Velocity

The principal purpose in such formulations is to obtain discontinuous approximations of **u** (with continuous **u**·**n** at the interelement faces) so as to apply upwinding techniques "à la LeSaint" as described in §2-5.[68] Due to the discontinuities in **u**, the approximation of the viscous term requires a "mixed" formulation, i.e. the introduction of an auxilliary variable: the vorticity $\omega = \text{curl} \, \mathbf{u}$ [69] (Girault and Raviart (1979)) or the deviatoric stress tensor $\sigma_{ij} = (u_{i,j} + u_{j,i})/2$ (Johnson (1978)). We begin with the first case; we note that Stokes equations can be written as:

$$\begin{cases} \text{a)} & \nu \, \text{curl} \, \omega + \text{grad} \, p = \mathbf{f} \\ \text{b)} & \omega - \text{curl} \, \mathbf{u} = 0 \\ \text{c)} & \text{div} \, \mathbf{u} = 0 \\ \text{d)} & \mathbf{u}|_\Gamma = 0 \end{cases} \qquad (136)$$

---

[68] See Chapter 4 for the application to Navier Stokes.

[69] $\omega$ is a vector in a 3-dimensional problem; we shall consider only the 2-dimensional problem.

It may be seen that the proper[70] variational space in which the velocity should be looked for consists of velocity fields with square integrable divergence, i.e. with Raviart's notations:

$$H_0(\text{div},\Omega)=\{v\in L^2(\Omega) \text{ such that: div}\,v=0, \, v\cdot n|_\Gamma=0\} \tag{137}$$

with this definition the weak formulation of (136) is as follows:

$$
\begin{cases}
\text{Find } \omega\in H^1(\Omega),\, u\in H_0(\text{div},\Omega),\, p\in L^2(\Omega): \\[2mm]
\text{a)} \quad \nu\int_\Omega v\cdot\mathbf{curl}\,\omega\,dx - \int_\Omega p\,\text{div}\,v\,dx = \int_\Omega f\cdot v\,dx, \quad \forall\, v\in H_0(\text{div},\Omega) \\[2mm]
\text{b)} \quad \int_\Omega \omega\theta\,dx - \int_\Omega u\cdot\mathbf{curl}\,\theta\,dx = 0 \qquad\qquad \forall\,\theta\in H^1(\Omega) \\[2mm]
\text{c)} \quad \int_\Omega q\,\text{div}\,u\,dx = 0 \qquad\qquad\qquad\qquad \forall\, q\in L^2(\Omega)
\end{cases}
\tag{138}
$$

**Remarks.**

(i) The boundary conditions on $u$ are not all satisfied in the same manner: "$u\cdot n=0$" is strongly imposed by the definition (137) of $H_0(\text{div},\Omega)$; the tangential boundary condition "$u\cdot t=0$" is implied by (138a), through the Green's formula.

(ii) Other boundary conditions can be introduced instead of (136c); for instance: the tangential velocity and the pressure on a part of the boundary; the tangential and normal components of the velocity along the rest of $\Gamma$:

$$
\begin{aligned}
u\cdot t=a, \qquad p=\pi_1 \quad \text{on } \Gamma_1 \\
u\cdot t=a, \qquad u\cdot n=b_2 \quad \text{on } \Gamma_2 = \Gamma\backslash\Gamma_1
\end{aligned}
\tag{139}
$$

Then the definition of $H_0(\text{div},\Omega)$ and the weak formulation are modified as follows:

$$H_0(\text{div},\Omega)=\{v\in L^2(\Omega) \text{ such that: div}\,v=0, \, v\cdot n|_\Gamma=0\}$$

$$\nu\int_\Omega v\cdot\mathbf{curl}\,\omega\,dx - \int_\Omega p\,\text{div}\,v\,dx$$

$$= \int_\Omega fv\,dx - \int_{\Gamma_1} \pi_1 v\cdot n\,ds, \qquad \forall\, v\in H_0(\text{div},\Omega)$$

$$\int_\Omega \omega\theta\,dx - \int_\Omega u\cdot\mathbf{curl}\,\theta\,dx = \int_\Gamma a\theta\,ds, \qquad \forall\,\theta\in H^1(\Omega)$$

$$\int q\,\text{div}\,u\,dx=0, \qquad \forall\, q\in L^2(\Omega)$$

$$u\cdot n=b_2 \quad \text{on } \Gamma_2$$

---

[70] That with just the desired continuity requirements on $u\cdot n$.

In the rest of this section we will consider exclusively homogeneous boundary condition (136d), for the sake of simplicity.

Let us consider now how the discrete equations will be obtained from (138), given a triangular, or rectangular, mesh, we will approximate $u$, $p$, $\omega$ by piecewise polynomials. The corresponding finite elements are founded on the following lemma (Thomas (1977), Girault and Raviart (1979):

> For a piecewise polynomial $v_h$, a necessary and sufficient condition for: $\operatorname{div} v_h \in L^2(\Omega)$ is the continuity of normal fluxes $v_h \cdot n$ at the interelement boundaries:

$$v_h^K \cdot n_K + v_h^{K'} \cdot n_{K'} = 0 \quad \text{on } K \cap K' \tag{140}$$

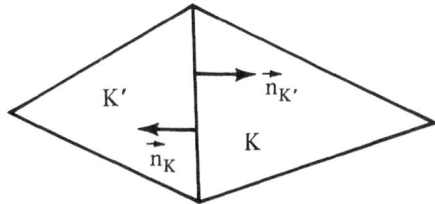

Finite elements satisfying (140) were analyzed by Thomas (1977) and implemented by Jaffre (1979); we saw the lowest degree element in §1-5; the degrees of freedom of this element are the values of normal velocity at mid points of each element-side.

A higher order element described by Thomas or Jaffre uses incomplete polynomials in $P2$ or $Q2$; the degrees of freedom for $u_h$ are in this case: i) the values of $u_h \cdot n$ at two points per side (it is convenient to chose the Gauss integration points along each side); ii) the values of the element integrals: $\int_K u_{h1}\, dx$, $\int u_{h2}\, dx$ (that is, 8 nodes per triangle, 10 nodes per quadrangle). Once the approximation of $u$ is chosen, the approximation of $p$ and $\omega$ is not arbitrary: (138) is a saddle point problem and a condition of Brezzi's type needs to be satisfied, in order to assure the unicity of the discrete solutions $p_h$, $\omega_h$. The following choices can be proved to be compatible with the last mentioned approximation of $u$ (8 node triangle, 10 node quadrangle)[71]:

$p_h$: discontinuous $P1$ or $Q1$
$\omega_h$: continuous piecewise $P2$ or $Q2$.

**Remark.** Since the pressure field is discontinuous, the incompressibility condition can be dealt with via the penalty formulation as in §3-2-5.

The above formulations are valid in two or three dimensional problems as well.

The implementation of the method can be further simplified in the case of *two* dimensional problems via the use of a *stream function*; indeed, for

---

[71] See also El Manouzi (1979) for the implementation.

$\mathbf{u} \in H_0(\mathrm{div}, \Omega)$: $\mathrm{div}\,\mathbf{u} = 0$ is equivalent to $\mathbf{u} = \mathrm{curl}\,\psi = \{\psi_{,2}, -\psi_{,1}\}$ with $\psi$ lying in the following space $\Phi = \Phi(\Omega)$:

$\Phi = H_0^1(\Omega)$,   if $\Gamma$ is connected;

if $\Gamma = \Gamma_0$ (: exterior boundary[72]) $\cup \Gamma_1 \cup \Gamma_2 \dots \cup \Gamma_m$,

$\Phi = \{\phi \in H^1(\Omega) \text{ such that:}$

$\phi|_{\Gamma_0} = 0$

$\phi|_{\Gamma_i} = $ some arbitrary constant, $i > 0$ (i.e. $\nabla \phi \times \mathbf{n} = 0$ on $\Gamma_i$)$\}$

Then an equivalent formulation to (138) is:

Find $\psi \in \Phi$, $\omega \in H^1(\Omega)$ such that:

$$\nu \int_\Omega \mathbf{grad}\,\omega \cdot \mathbf{grad}\,\phi\, dx = \int_\Omega \mathbf{f} \cdot \mathbf{curl}\,\phi\, dx, \qquad \forall\, \phi \in \Phi$$

$$\int_\Omega \omega\theta\, dx - \int_\Omega \mathbf{curl}\,\psi \cdot \mathbf{curl}\,\theta\, dx = 0, \qquad \forall\, \theta \in H^1(\Omega)$$

$$\mathbf{u} = \mathbf{curl}\,\psi$$

Standard Lagrange-type finite elements can be used to discretize (140), e.g. piecewise $P2$ (continuous; i.e. the 6 node element) for $\omega$ and $\psi$, yielding the following error estimates:

$$\left( \int_\Omega |u - u_h|^2\, dx \right)^{1/2} = O(h)$$

$$\left( \int_\Omega |\omega - \omega_h|^2\, dx \right)^{1/2} = O(h)$$

Similar methods, where instead of $\omega$ the auxilliary variable is either $\partial^2\phi / \partial x_i \partial x_j = $ tensor of second derivatives of the stream function (Figueroa (1979)); or $\Sigma_{ij} = 1/2(\partial u_i/\partial x_j + \partial u_j/\partial x_i) = $ deviatoric stress tensor (Johnson (1978), El Manouzi (1979)). For instance the latter reference analyzes the approximation of the following weak formulation[73]:

$$\mathbf{u} \in H(\mathrm{div}, \Omega) = \left\{ \mathbf{v} \in L^2(\Omega) \right\}^2 \quad \text{such that} \quad \mathrm{div}\,\mathbf{v} \in L^2(\Omega)$$

$$(\varepsilon_{ij}) = \tilde{\varepsilon} \in X = H(\mathrm{div}, \Omega) \times H(\mathrm{div}(\Omega))$$

$$p \in H^1(\Omega)$$

---

[72] Supporting the conditions "at infinity" in the case of the flow around an airfoil (for instance).

[73] Other formulations are also analyzed; the present one is adapted to the introduction of upwinding in the approximation of Navier Stokes equations.

$$\sum_{i,j=1}^{2} \int_{\Omega} \varepsilon_{ij}\eta_{ij}\,dx + \int_{\Omega}\left(\sum_{i,j} u_i\frac{\partial\eta_{ij}}{\partial x_j} - \sum_i u_i\frac{\partial q}{\partial x_i}\right)dx = 0,$$

$$\forall\,\tilde{\eta}\in X, \quad \forall\,q\in H^1(\Omega)$$

$$\int_{\Omega}\left(\sum_{i,j=1}^{2} v_i\frac{\partial\varepsilon_{ij}}{\partial x_j} - \sum_{i=1}^{2} v_i\frac{\partial p}{\partial x_i}\right)dx = 0 \quad \forall\,v\in H(\text{div},\Omega), \, v\cdot n = 0 \quad \text{on } \Gamma.$$

$u\cdot n = 0$   on $\Gamma$.

The reasons for the interest in such formulations are twofold:

(i) The accuracy is one order of magnitude better than with an $\omega - \psi$, or $\omega - u - p$, formulation;
(ii) The stress tensor has a fundamental physical meaning for which boundary conditions may be required, e.g. along free surfaces or at exit boundaries.

However a relatively heavy price has to paid in the number of variables; for instance the lowest order method described by El Manouzi (1979) uses incomplete polynomials of degree 1 for $u$ and $\varepsilon$,[74] conforming $P1$ triangles for $p$; the resulting number of degrees of freedom is of order: $5\times$Number of Triangles, for an order of accuracy:

$$\left(\int_{\Omega}|u - u_h|^2\,dx\right)^{1/2} = O(h)$$

$$\left(\int_{\Omega}|\varepsilon - \varepsilon_h|^2\,dx\right)^{1/2} = O(h)$$

$$\left(\int_{\Omega}|p - p_h|^2\,dx\right)^{1/2} = O(h)$$

## 3.5. Vorticity Stream-Function Formulation: Decompositions of the Biharmonic Problem

In two dimensions of space, the introduction of the vorticity $\omega = \text{curl}\,u$ and of the stream function $\psi$ transforms the Stokes problem (95) into the following Poisson's equations:

a)  $-\Delta\psi = \omega$

b)  $-\Delta\omega = -\text{curl}\,f \equiv +g$                                    (141)

c)  $\psi|_\Gamma = 0$,[75]     $\dfrac{\partial\psi}{\nu} = 0$

---

[74] So that the normal components of $u$ and of each column of tensor $\varepsilon$ are continuous across interelement boundaries;

[75] This is valid for a *connected* boundary; otherwise $\psi$ is an unknown constant on each obstacle in the flow: $\psi\in\Phi$ with the notations of previous section §3-4.

Boundary conditions may be prescribed in some cases for $\omega$: e.g. at upstream and outflow boundaries, assuming a fully developed flow; but such conditions are not available at all parts of $T$ for a general problem. This difficulty has been thoroughly studied in finite difference context: for instance, near a wall—where the noslip condition applies for $\mathbf{u}$—the classical remedy is to use Taylor expansions of $\psi$ in the normal direction, together with the noslip condition, to obtain an approximate boundary condition in $\omega$; for instance:

$$\omega_0 = 2\frac{\psi_1 - \psi_0}{\partial x^2} + O(\partial x)$$

(Thom's formula: $\omega_0$ and $\psi_0$ are the values of $\omega$ and $\psi$ at the wall which is assumed parallel to $Oy$; $\psi_1$ is the value of $\psi$ one mesh point away from the wall. These methods are reviewed in Roache (1972, chapter III-C, pp. 140–160).[76] Although the straightforward[77] finite element extension of such techniques has been tried (Datta and Strauss (1976), Taylor and Hood (1973), Ikegawa (1979)), caution is required in the application of such a technique: no informations are available as to the *reliability* of such a technique in general configurations.

We briefly describe now three variants of one decomposition method, using standard finite elements of Lagrange type, e.g. $P2$ or $P1$ triangles, $Q2$ or $Q1$ quadrangles, etc. We use the same notations as in §3-3-3 (for a given triangulation and given $k = 1, 2, \ldots$).

$W_h$ = set of continuous piecewise polynomials[78] of degree $\leq k$ (see chapter 1): for each element $K$ and $\omega \in W_h$: $\omega|_K \in P_k$ if $K$ is a triangle, $\omega|_K \in Q_K$ if $K$ is a quadrangle.

$W_{h0}$ = set of functions in $W_h$ vanishing on $\Gamma$

$\mathfrak{M}_h$ = subspace of $W_h$ generated by basis functions associated to boundary nodes:

$$W_h = W_{h0} \oplus \mathfrak{M}_h$$

$\omega_h \in W_h$: discrete vorticity
$\lambda_h \in \mathfrak{M}_h$: (unknown) value on the boundary, of the (discrete) vorticity $\omega_h$
$\psi_h \in W_{h0}$: discrete stream function.

---

[76] Recently by Gupta and Manohar (1979).

[77] i.e., writing difference-like boundary conditions such as Thom's formula, connecting the values of $\omega$ and of $\psi$ at 2 neighbor points in the normal direction at the boundary.

[78] in the case of a curved boundary, isoparametric elements should be used.

Let us write the discrete analogues of (141):

$$\begin{cases} \omega_h - \lambda_h \in W_{h0} \\ \displaystyle\int_\Omega \mathbf{grad}\,\omega_h \cdot \mathbf{grad}\,\phi \, dx = \int_\Omega g\phi \, dx, \qquad \forall \phi \in W_{h0} \end{cases} \tag{142}$$

$$\begin{cases} \psi_h \in W_{h0} \\ \displaystyle\int_\Omega \mathbf{grad}\,\psi_h \cdot \mathbf{grad}\,\phi \, dx = + \int_\Omega \omega_h \phi \, dx, \qquad \forall \phi \in W_{h0} \end{cases} \tag{143}$$

The first boundary condition in (141c) is taken into account by $\psi_h \in W_{h0}$ (i.e. freezing to 0 the values of $\psi_h$ on the boundary). We need discrete equations to represent the other condition $\partial\psi/\partial n = 0$; if $\psi$ and $\mu$ are smooth functions, with $\psi$ solution of (141) we have by the Green's formula:

$$0 = \int_\Gamma \frac{\partial \psi}{\partial n} \mu \, ds = \int_\Omega (\mathbf{grad}\,\psi \cdot \mathbf{grad}\,\mu - \omega\mu) \, fx$$

Therefore the discrete version of $\partial\psi/\partial n = 0$ is:

$$\int_\Omega \mathbf{grad}\,\psi_h \cdot \mathbf{grad}\,\mu_h \, dx = \int_\Omega \omega_h \mu_h \, dx, \qquad \forall \mu_h \in \mathfrak{M}_h \tag{144}$$

*First method* (Campion-Renson and Crochet (1978), Ikenouchi and Kimura (1974))

Putting together (142), (143) and (144) we get a coupled system in $\omega_h, \psi_h$:

$$\boxed{\begin{aligned} &\omega_h \in W_h, \psi_h \in W_{h0} \\[4pt] &\int_\Omega \mathbf{grad}\,\omega_h \cdot \mathbf{grad}\,\phi \, dx = \int_\Omega g\phi \, dx, \qquad \forall \phi \in W_{h0} \\[4pt] &\int_\Omega (\mathbf{grad}\,\psi_h \cdot \mathbf{grad}\,\phi - \omega_h \phi) \, dx = 0, \qquad \forall \phi \in W_h \end{aligned}} \tag{145}$$

(see also Axelsson and Gustafsson (1979) for iterative solution of (145)).

*Second method:* "dual iterative method" (Ciarlet and Glowinski (1975), Glowinski (1973), Ciarlet and Raviart (1974), Bossavit (1971), Bourgat (1976)).

The above system of equations is equivalent to a saddle point problem, where $\lambda_h = \omega_h|_\Gamma$ is the Lagrange multiplier associated to the constraint (144), i.e. ("$\partial$"/$\partial n$)$\psi_h = 0$.

Hence the following algorithm of Uzawa type, with parameter $\rho$:

Step 0: initialize $\lambda_h$, e.g. $\lambda_h = 0$

Step 1: compute $\omega_h$ solution of (142)

Step 2: compute $\psi_h$ solution of (143)

Step 3: compute $\bar{\lambda}_h \in \mathfrak{M}_h$ such that:

$$\forall \mu \in \mathfrak{M}_h: \int_\Gamma \bar{\lambda}_h \mu \, ds = \int_\Gamma \lambda_h \mu \, ds + \rho \int_\Omega (\text{grad } \psi_h \cdot \text{grad } \mu - \omega_h \mu) \, dx$$

Step 4: if the algorithm has not yet converged, set $\lambda_h = \bar{\lambda}_h$
and go to step 1.

$$(146)$$

**Remarks.**

(i) the choice of $\rho$ is empirical for a general domain, although bounds for $\rho$ can be given in the case of domains with simple geometries; e.g.: $0 < \rho < 2M$ in the case of square $\Omega$; $0 < \rho < 8$ when $\Omega$ is a disk. From Bourgat (1976), about 20 iterations are enough for convergence in a typical test problem;

(ii) steps 1 and 2 of (146) involve discrete Poisson's equations, to be solved e.g. by SOR or Cholesky factorization (the matrix is invariant during the iterations);

(iii) step 3 should be made *explicit* through a proper quadrature formula over $\Gamma$ (in fact any inner product in $\mathfrak{M}_h$ can yield a convergent algorithm);

(iv) for further details on the implementation and numerical results see Bourgat (1976). For related ideas see also Bonnet and Perronnet (1973).

*Third method* (Glowinski and Pironneau (1977)

This method is similar to the decomposition of Stokes problem, §3-3-3: here (142), (143), (144) yield a boundary equation for $\lambda_h$; this equation is formed through the solution of a "cascade" of discrete Poisson's equations.

The matrix of the $\lambda_h$-equation is formed just in the same manner in §3-3-3 and can be inverted either by the Cholesky factorization or by a conjugate gradient method: it results from the numerical experiments of M. Vidrascu (1978) that the latter is about twice faster and requires less core storage (by a factor 3).

**Remarks.** Other types of approximation of the biharmonic equations;

(i) for non standard finite element ("mixed" or "hybrid") see Brezzi and Marini (1975), Brezzi and Raviart (1976), Brezzi *et al.* (1979).

(ii) for the use of conforming methods (with $C^1$-regularity such as 21 node Argyris' triangle), or non conforming methods, see e.g. Ciarlet (1975, 1978).

(iii) in 3 dimensions of space, the introduction of the (vector valued stream function $\psi$ seems to be of little use, since anyway the condition div $\psi = 0$ needs to be imposed in order to insure the unicity to $\psi$ (see Bernardi (1979)).

# 4. Navier-Stokes Equations: Accuracy Assessments and Numerical Results

The goals of this chapter are: i) to comment the weak formulations of non linear equations using the methods of chapter 3 (possibly combined with the upwinding techniques of chapter 2); ii) to review some theoretical and numerical results available in the mathematical and engineering literature, regarding the accuracy of these methods. (The numerical procedures for the solution of the linear and non linear related systems are briefly reviewed in chapter 5).

When the solution of the mathematical problem is unique (i.e. at moderate Reynolds number, see e.g. Temam (1977a), Girault and Raviart (1979)) the methods of chapter 3 are proved to converge and the error estimates have the same order of accuracy as in the linear case.

## 4.1. Remarks on the Formulation

The equations of notion to be solved are: (together with appropriate boundary conditions):

*Newton's law:*

$$\rho\left(\frac{\partial u_i}{\partial t}+(u_i u_j)_{,j}\right)-\sigma_{ij,j}=\rho f_i$$

(147a)

($\rho$=fluid density)

*Mass conservation:*

$$u_{j,j}=0, \rho=\text{constant}$$

(147b)

*Newtonian constitutive law:*

$$\sigma_{ij}=-p\delta_{ij}+\mu(u_{i,j}+u_{j,i})$$

(147c)

Using the incompressibility conditions the acceleration terms in (147a) become:

$$\rho\left(\frac{\partial u_i}{\partial t}+u_j u_{i,j}\right)$$

(148a)

and the stress term:

$$\sigma_{ij,j} = -p_{,i} + \mu u_{i,jj} \tag{148b}$$

so that (147a) can be simplified to the classical form $(\nu = \mu/\rho)$:

$$\frac{\partial u_i}{\partial t} - \nu \Delta u_i + u_j u_{i,j} + (p/\rho)_{,i} = f_i \tag{149}$$

A lot of the finite element methods for Navier Stokes equations make use of approximate velocity with small by *non zero*, divergence[79]; so that the discrete equations driven either from (147)[80] or from (149) should not be strictly equivalent.[81] The equations of primary physical interest should be (147), yet a significant amount of finite element solutions of Navier Stokes equations use (149).

As far as the stress tensor is concerned there is a consensus of many authors to leave $\sigma_{ij}$ as it is in (147c) instead of replacing it by (148b): e.g. Hood and Taylor (1973), Hughes et al. (1979), Nickell et al. (1974).

As to the non linear term its formulation is discussed by Hutton et al. (1980), Hughes et al. (1979): the latter mention that in their numerical experiments with $Q2$ (nine nodes biquadratic) elements for **u**, and discontinuous $Q1$ (four nodes bilinear) elements for $p$, no discernable difference was observed when (148a) was used in place of the conservative form. The same authors also used without damage to the accuracy, the following form of the acceleration term:

$$\frac{\partial u_i}{\partial t} + \frac{1}{2}\left( u_j u_{i,j} + (u_i u_j)_{,j} \right)$$

which results in the following weak formulation proposed by Temam (1968) and useful for the *stability* of the numerical scheme:

$$\int_\Omega \left( \frac{\partial u_i}{\partial t} v_i + \frac{1}{2}(u_j u_{i,j} v_i - u_j u_i v_{i,j}) \right.$$

$$\left. + \frac{\nu}{2}(u_{i,j} + u_{j,i})(v_{i,j} + v_{j,i}) - p v_{j,j} - f_i v_i \right) dx = 0, \qquad \forall\, \mathbf{v} = \{v_1, v_2\}$$

It should be useful to know whether such conclusions on the practical equivalence of conservative and non conservative forms, also hold for other finite element approximations, in particular when the approximate pressure is continuous.

---

[79] In the limit of vanishing meshsize, $\mathrm{div}\, \mathbf{u}_h \to 0$; but on the computer you work with a finite mesh.

[80] By analogy with finite differences, the formulation driven from (147) might be named "conservative."

[81] Of course the finite element methods for the solution of (149) may be adapted without difficulty, but with variable efficiency, for the solution of (147). Note that the transportive form (148a) is better suited for the introduction of upwinding.

With these restrictions in mind there is no difficulty to write, at least formally, the weak formulations corresponding to (147) or (149): knowing the formulation of Stokes problem, just add to $-f_i$ (the body force) the acceleration term.

We note that the use of upwinding techniques of chapter 2 sets no difficulty when (148a) or (149) is used ((149) is a transport equation for the velocity itself); for instance with a semi implicit time discretization of (149):

$$\frac{\mathbf{u}^{m+1}-\mathbf{u}^m}{\partial t}+(\mathbf{u}^m\cdot\nabla)\mathbf{u}^{m+1}-\nu\Delta\mathbf{u}^{m+1}+\nabla\left(p^{m+1}/\rho\right)=\mathbf{f}^{m+1},$$

$\mathbf{u}^m$ is known when we have to compute $\mathbf{u}^{m+1}$; so that at each time step we solve a stationary advection/diffusion problem, with known velocity field as studied in chapter 2.

## 4.2. A Review of the Different Methods

Now we review the methods described in chapter 3, indicate the possibilities of upwinding, and give the references for convergence proofs and numerical results (the wall-driven square cavity problem is considered to be one of the significant possible tests). Some numerical results are shown in §4-3.

### 4.2.1. Velocity–Pressure Formulations: Discontinuous Approximations of the Pressure

Convergence proofs: Jamet and Raviart (1973), Temam (1977).

$\mathbf{u}_h$: non conformity $P1$; $p_h$: discontinuous $P0$ (i.e. piecewise constant) (§1-4-5, §3-2-1).
Numerical results: Thomasset (1976, 1977, 1980) (using the zero divergence basis).
For the square cavity problem, the results compare fairly well to the solution by accurate finite differences, for Reynolds numbers up to 100 (with a comparatively cheap number of variables); at $R=400$, the method failed to give the correct solution (see Figure 33; of course a still finer mesh would— hopefully—yield the correct answer).
$\mathbf{u}_h$: $P2$, $p_h$: $P0$ (§3-2-3)
Convergence proof and numerical results in Fortin (1972): results are shown for the square cavity problem at $R\leq100$. (To the author's knowledge, solution at higher Reynolds numbers was not tried with this element).
$\mathbf{u}_h$: $Q2$, $p_h$: discontinuous $Q1$ (§3-2-5)
Convergence proof and numerical results in Engelman (1979), Bercovier and Engelman (1979): satisfying solutions are obtained in the square cavity problem at $R=0,100,400,1000$. Also Hughes et al. (1979) tested their up-winding procedure (§2-3) on this finite element method for Navier Stokes

equations. The effects of upwinding were most clearly seen in the problem of flow past a step as shown by these authors: using a coarse mesh they get non physical wiggles which are smoothed out when the upwinding is introduced.

### 4.2.2. Velocity–Pressure Formulations: Continuous Approximations of the Pressure

(Cf. §3-3)
Convergence proof in Letallec (1978) (who also gives some numerical results).

This is probably the most popular formulation among aerodynamicists, possibly because of the feeling that an approximation with continuous functions is more attractive than discontinuous functions: yet there is no argument in favour of approximation by continuous functions, as far as the accuracy is concerned. Numerical results can be found in Hood and Taylor (1973), Nickell, Tanner and Caswell (1975); Periaux (1978), Glowinski *et al.* (1979, 1980) use a non linear conjugate gradient algorithm (see chapter 5), so that the non linear equations are solved with a succession of Stokes problems, where the decomposition technique of §3-3-3 can be used: numerical results obtained by J. Periaux with this technique for an engineering problem is given in §4-3-2; the algorithm is equally efficient for 3-D problems.

Benque *et al.* (1980) used a splitting technique (see chapter 5) so that the convective and diffusive terms are treated in successive steps: the first uses the method of characteristics (§2-6) and the latter step (which a Stokes equation) uses the same decomposition technique described in §3-3-3.

A comparison of several schemes was performed by Huyakorn et al. (1978), in particular for the problem of flow through a 2-D channel with a sudden enlargement: from this results, the "$Q2+Q1$" and "$P2+P1$" elements give a better accuracy than the 8-nodes "serendip" quadrangle (for $u$) of Zienckiewicz (1977) (with 4-nodes $Q1$ approximation of $p$).

### 4.2.3. Vorticity–Pressure–Velocity Formulations: Discontinuous Approximations of Pressure and Velocity

(Cf. §3-4)
Convergence proofs: Girault and Raviart (1979).

Such approximations make use of Lesaint's upwinding technique (§2-5) so that the method is stable at high Reynolds numbers. Up to now these methods have received little attention from engineers, probably because: i) of the non standard elements[82] involved when the stream function is not used; ii) the comparatively large number of variables: in this respect the use of recent techniques for the solution of linear systems (like incomplete Cholesky factorization, see chapter 5) can bring a renewal of interest in those methods.

---

[82] Introduced by Raviart-Thomas, see §3-4 and 1-5.

Let us briefly describe the discrete formulation of stationary Navier Stokes in the 2-D case, when the vorticity and stream function can be taken as dependent variables. To fix ideas we use $P2$ triangles (or $Q2$ quadrangles) for the approximation of $\omega$ and $\psi$.

The discrete equations are (with the notations of (140), §3-4):

$W_h$: set of piecewise quadratic ($P2$ or $Q2$) functions in $H^1(\Omega)$

$\Phi_h$: set of piecewise quadratic ($P2$ or $Q2$) functions in $\Phi$

Find $\psi_h \in \Phi_h$, $\omega_h \in W_h$ such that

$$\nu \int_\Omega \text{grad } \omega_h \cdot \text{grad } \phi \, dx + b(u_h, u_h, \text{curl } \phi) = \int_\Omega \mathbf{f} \cdot \text{curl } \phi \, dx, \qquad \forall \, \phi \in \Phi_h$$

(150)

$$\int_\Omega \omega_h \phi \, dx - \int_\Omega \text{curl } \psi_h \, \text{curl } \phi \, dx, \qquad \forall \, \phi \in W_h \tag{151}$$

$$u_h = \text{curl } \psi_h$$

$b(u_h, u_h, \text{curl } \phi)$ is the upwind approximation of the advection term: $\int_\Omega \text{"}(u_h \cdot \nabla)\text{"} \, u_h \, \text{curl } \phi \, dx$; more precisely, with $u_h = \{u_1, u_2\}$:

$$b(v_h, u_h, w_h) = \sum_T \left( -\int_T v_j u_{i,j} w_i \, dx + \int_{\partial T} (\mathbf{v} \cdot \mathbf{n}_T)(\alpha u_h^+ + (1-\alpha)u_h^-) \cdot w_h \, ds \right)$$

(152)

(summation over the elements, $\partial T$ = boundary of element $T$, $\mathbf{n}_T$ = outward unit vector on $\partial T$).

In (151), $u_h^+$ and $u_h^-$ are respectively the upstream and downstream values of the velocity according the sign of $u_h \cdot \mathbf{n}_T$ (which by the way is continuous across $\partial T$):

When $u_h \cdot \mathbf{n}_T > 0$, $u_h^+ =$ value of $u_h$ computed within the element $T$, and $u_h^- =$ value of $u_h$ computed within the neighbouring (downstream) element $T'$;

Vice versa when $u_h \cdot \mathbf{n}_T < 0$ (see Figure 29)

$\alpha$ is an upwinding parameter ($\frac{1}{2} \le \alpha \le 1$; Fortin (1976) has shown that the

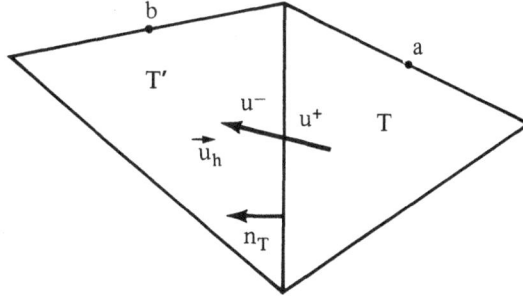

Fig. 29

numerical dissipation of the scheme is proportional to $(2\alpha - 1)$); this $\alpha$ does not need to be constant; it might be equal to $1/2$ in regions where diffusion effects dominate advection, and nearer to 1 elsewhere (just as e.g. in Hughes' scheme). However this possibility has not been explored.

Two dimensional numerical results in $\omega - \psi$ formulation (on the square wall driven cavity, see §4-3-1) shown by Thomasset (1977), Fortin and Thomasset[83] (1979) (using $P2$ triangles); by Benazeth (1977), Paillard (1979) (using $Q2$ quadrangles). Note that quadrangles give better results than triangles on this example.

Fortin and Thomasset solved the transient problem with a semi-explicit[84] time-marching scheme; Benazeth and Paillard considered the solution of the stationary problem through a successive approximation iterative scheme as follows: ($m$: iteration counter)

$$\nu \int_{\Omega} \operatorname{grad} \omega_h^{m+1} \operatorname{grad} \phi \, dx + b\left(u_h^m, u_h^{m+1}, \operatorname{curl} \phi\right) = \int_{\Omega} \mathbf{f} \cdot \operatorname{curl} \phi \, dx, \forall \, \phi \in \Phi_h$$

$$\int_{\Omega} \left(\omega_h^{m+1}\phi - \operatorname{curl} \psi_h^{m+1} \cdot \operatorname{curl} \phi\right) dx = 0, \forall \, \phi \in W_h \qquad (153)$$

$$\mathbf{u}_h^{m+1} = \operatorname{curl} \psi_h^{m+1}$$

Note that (153) is a large system with non symmetric, $m$ dependent, matrix (whose bandwidth is increased by the introduction of upwinding: in Figure 29, node $a$ and node $b$ will be connected); this was solved by a frontal method (Joly (1977)). Let us also notice that the non linear equations (153) are non differentiable[85] with respect to $\mathbf{u}_h$ so that a direct application of Newton-Raphson method to (153) is at least questionable.

Other related methods were also tested on the square cavity problem: Figueroa (1979) used as an auxilliary variable in place of $\omega$, the tensor of second derivatives $\psi_{,ij}$ of the stream function; from comparisons with an upwind finite difference code with $50 \times 50$ mesh, this seems to be more accurate than the $\omega$-$\psi$ formulation; El Manouzi (1979) used $\varepsilon_{ij} = \mu(u_{i,j} + u_{j,i})$ = viscous stress tensor as an auxilliary variable.

### 4.2.4. Vorticity–Stream-Function Formulation

(Cf. §3-5)

This is the classical $\omega$-$\psi$ formulation. Unlike the preceding method, the vorticity is related to the stream function through a Poisson's equation

---

[83] Also shown in this reference is the 2-D flow past a cylinder.

[84] Requiring the solution of two symmetric positive definite systems per iteration.

[85] But the non differentiable terms are small (see V. Girault, thesis, 1979).

$(R=1/\nu)$

$$-\Delta\psi=\omega$$

$$-\Delta\omega+R(\psi_{,2}\omega_{,1}-\psi_{,1}\omega_{,2})=\text{curl } f$$

$$+\text{boundary conditions }\left(\text{e.g. }\psi\Big|_{\Gamma}=0,\frac{\partial\psi}{\partial n}\Big|_{\Gamma}=0\right)$$

Numerical results for the formulation analogue to (142), (143) can be found in Campion-Renson and Crochet (1978), in the square cavity, for $R=$ 0, 100, 400; Ikegawa (1979) gives numerical results of his method for u modified cavity problem (the fluid is driven by a thin channel flow) and compares with physical experiments at $R\simeq 1400$.

The dual iterative method (146) was implemented for Navier Stokes equations by M. Borrel (1977), who shows again numerical results on the square cavity at $R\leq 400$, and for the problem of flow near the trailing edge of an airfoil ($R=400$); it is likely that the third method of §3-5 proposed by Glowinski and Pironneau (1977) would be more effective, although no numerical tests are available for the non linear case (these authors preferred to develop the analogue u, p formulation §3-3-3, which extends to 3-D problems).

## 4.3. Some Numerical Tests

### 4.3.1. The Square Wall Driven Cavity Flow

The problem definition is in Figure 30; the *stationary* problem is considered. It does not really show the full possibilities of finite element methods, since a large number of finite difference workers solved the problem since Burggraff (1966), e.g. Bozeman and Dalton (1973), Nallasamy and Krishna-Prasad (1977), Benjamin and Denny (1979), Dennis *et al.* (1979).

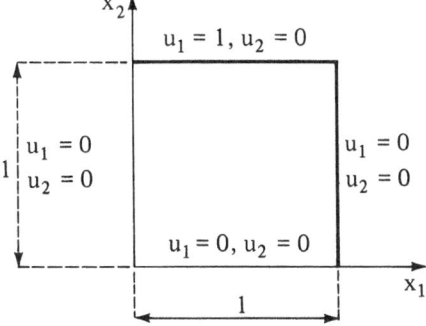

Fig. 30

Fig. 31  408 triangles

Yet precisely because of this large amount of references, this problem is felt as a simple test for a new code. We note that the finite difference papers give similar results for $R=$ Reynolds number $\leq 1000$, but that the results are controversial at higher Reynolds, e.g. $R \geq 10000$. For instance Nallasamy and Krishna-Prasad (1977) show numerical results for $R=10^2$ to $3 \times 10^4$, where the size and strength of secondary eddies[86] *decrease* in strength when $R$ increases from $10^3$; Benjamin and Denny (1979) attributed this effect to the use of *upwind* differences and found the contrary effect (increasing secondary eddies). See also de Vahl Davis and Mallinson (1976).

However this controversy is likely to be solved sometimes in the future so that the square cavity problem remains a significant test. It is most regretable that no standard form exists for the presentation of the results (for finite

---

[86] That is smaller counter clockwise eddies in the right hand left down corners.

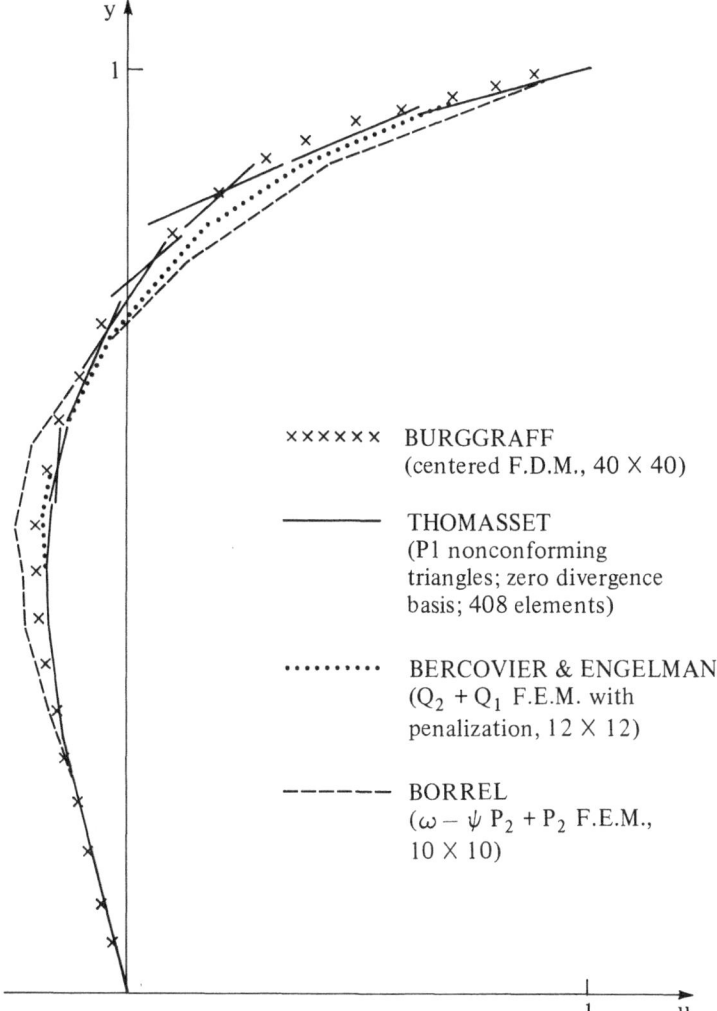

Fig. 32 Square wall driven cavity. $R = 100$. Profile of horizontal velocity along vertical centre line

difference as well as finite element papers) which makes the comparison more difficult.

The most common outputs are:

The shape of streamlines and velocity fields, which allows only qualitative comparisons; however comparisons of streamlines are possible with physical experiments (see e.g. Ikegawa (1979));

The values of maximum and minimum value of stream function (when available) and vorticity;

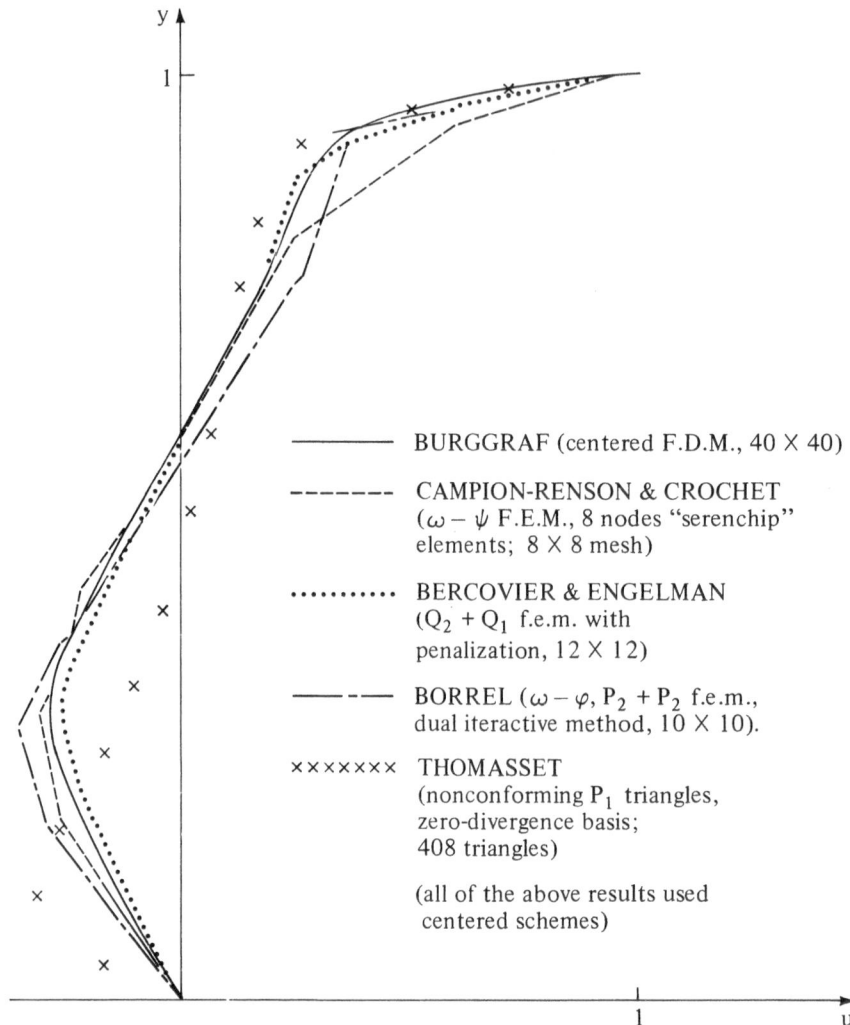

Fig. 33 Square wall driven cavity. $R=400$. Profile of horizontal velocity along vertical centre line

The location of the centre of main eddy;

The dimensions and strength of secondary eddies;

The variation along the centre line $x_1 = .5$, of the $u_1$ component of the velocity: we give these profiles for Reynolds numbers: 100, 400, 1000, 10000, and several f.e.m. and f.d. methods: see Figures 32 to 35.

Let us mention that these graphs are only indicative and that further tests should be performed in order to draw definite conclusions as regards accuracy: in fact, the methods ought to be compared for (approximately) equal

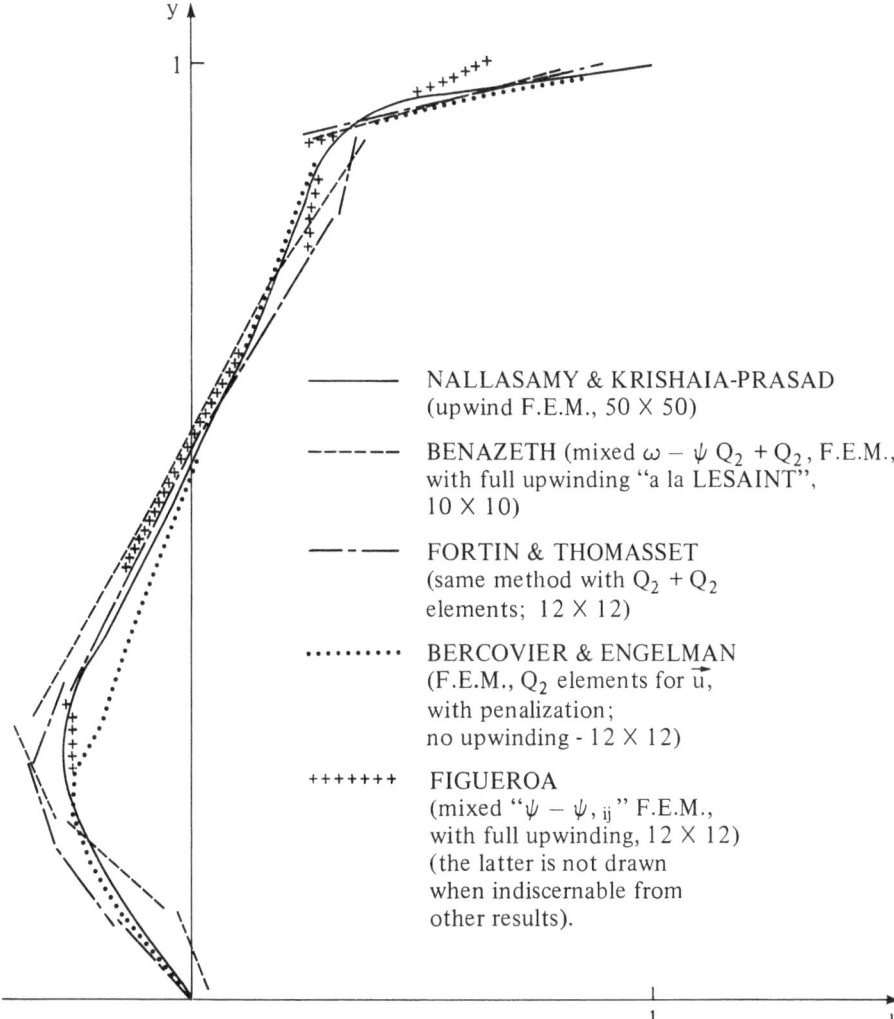

Fig. 34 Square wall driven cavity. $R=1000$. Profile of horizontal velocity along vertical centre line

numbers of unknowns. The meshes in use are regular meshes[87] made of small squares possibly divided into 2 triangles per square (the notation $12 \times 12$ means that $12^2 = 144$ small square elements are used).

Let us mention that Benazeth (1977, §4-2-3 of this course) measured the size of the computed secondary eddies and found that they decreased when $R$

---

[87]Except for the non conforming $P1$ element, for which a 408-triangles mesh was used (which however proved to be unsufficient at $R=400$ as indicated by Figure 33).

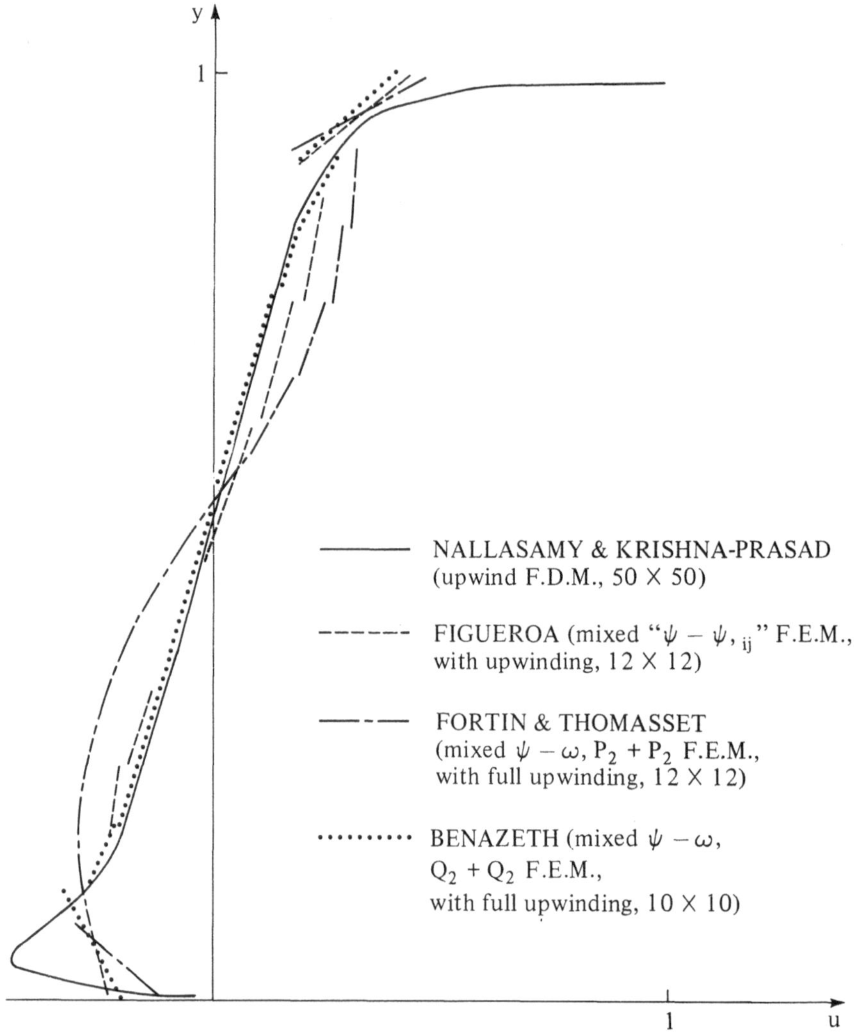

y

1

————————  NALLASAMY & KRISHNA-PRASAD
              (upwind F.D.M., 50 × 50)

- - - - - -  FIGUEROA (mixed "$\psi - \psi_{,ij}$" F.E.M.,
              with upwinding, 12 × 12)

— · — · —  FORTIN & THOMASSET
              (mixed $\psi - \omega$, $P_2 + P_2$ F.E.M.,
              with full upwinding, 12 × 12)

·············  BENAZETH (mixed $\psi - \omega$,
              $Q_2 + Q_2$ F.E.M.,
              with full upwinding, 10 × 10)

1                    u

Fig. 35 Square wall driven cavity. $R = 10000$. Profile of horizontal velocity along vertical centre line

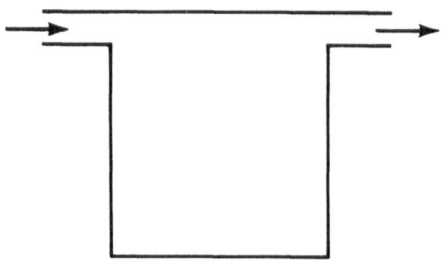

Fig. 36

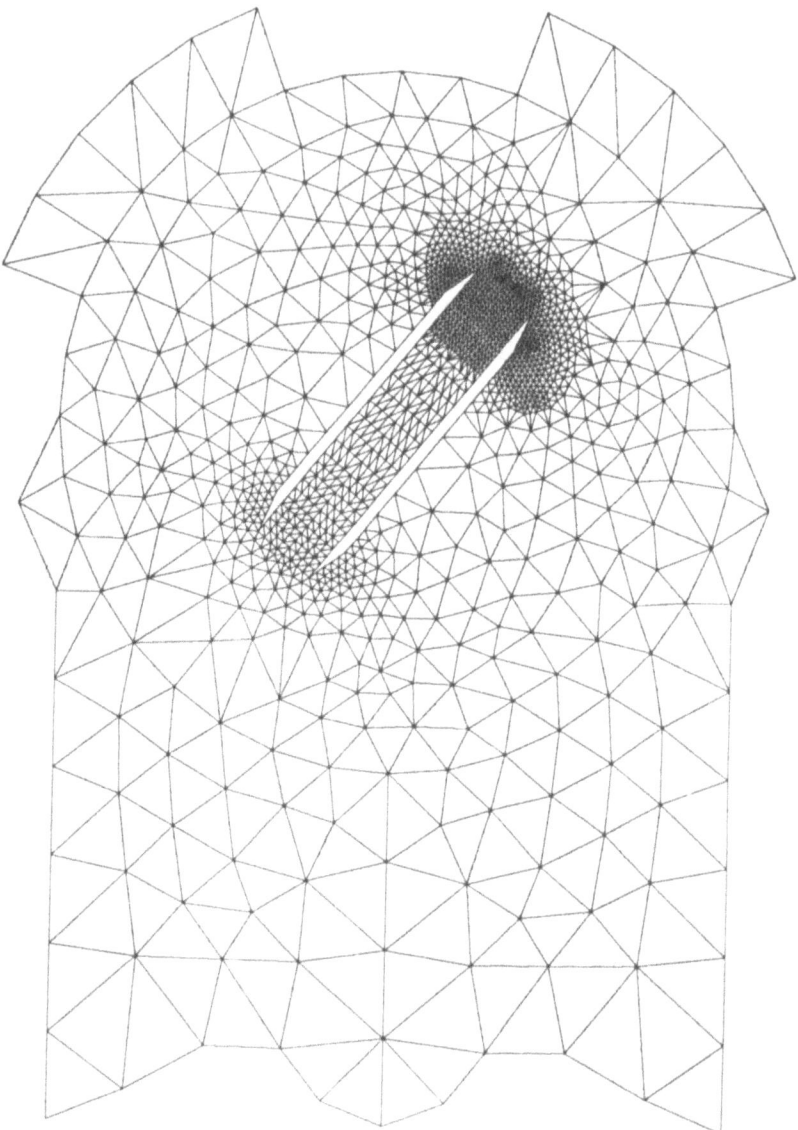

Fig. 37

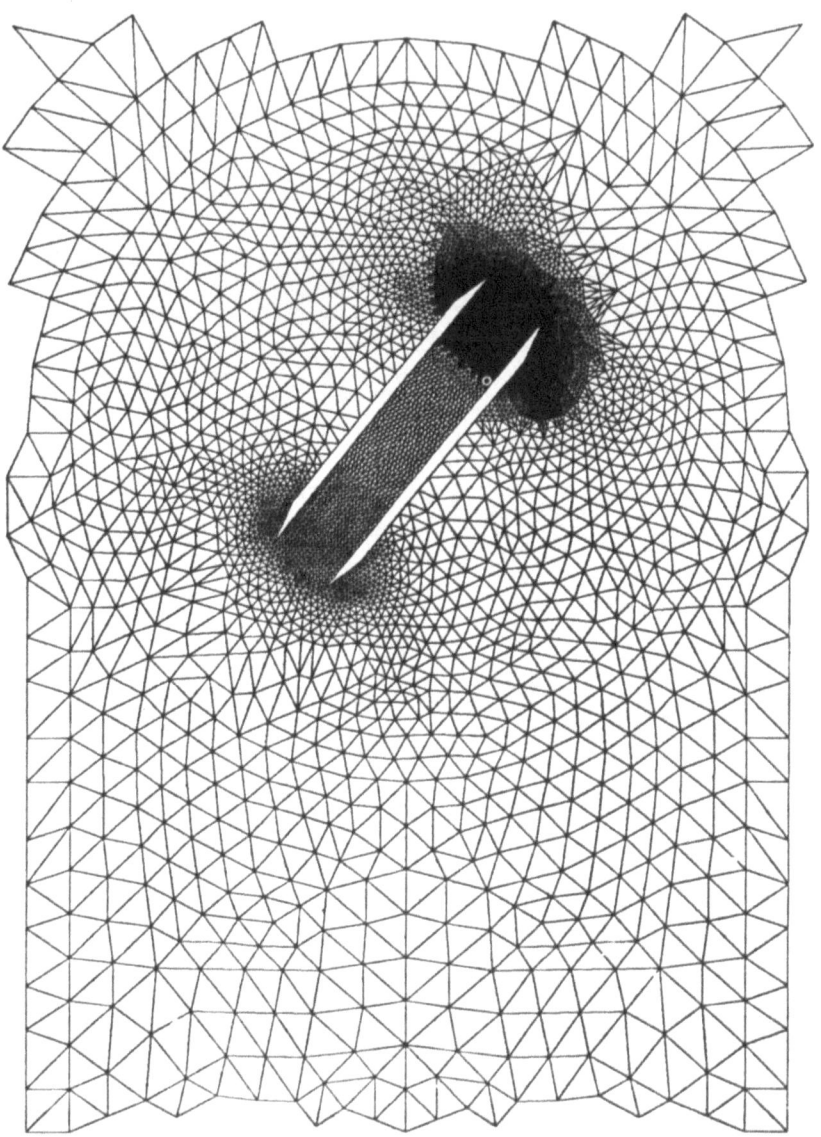

Fig. 38

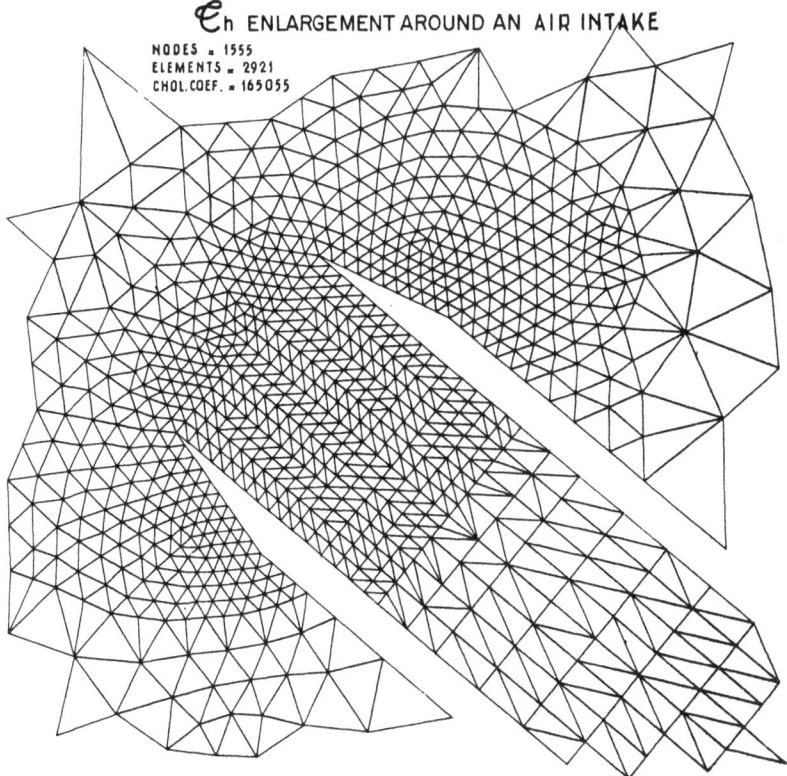

$\mathcal{C}_h$ ENLARGEMENT AROUND AN AIR INTAKE

NODES = 1555
ELEMENTS = 2921
CHOL.COEF. = 165055

Fig. 39

increased from $R = 1000$, as in Nallasamy and Krishna-Prasad (full *upwind* finite differences; it will be of interest to test whether this is really as spurious effect of upwinding, as Benjamin and Denny (1979) have suggested; note that Benazeth used a constant $\alpha$—upwinding parameter—throughout the calculation, which is likely to be non optimal).

Finally we mention that the results in Letallec (1978) show relatively poor results at $R = 1000$ with continuous approximation of $p$ ($P2$-$P1$ element, §3-3); this may be due to the singularity in the pressure field in the top corners, which is poorly represented by $P1$ conforming elements. A modified square cavity problem with at least the same physical interest, could be studied in order to obviate this numerical difficulty; namely to replace the moving wall by a thin channel (as Bercovier and Engelman (1979) did to compute a solution at $R = 2000$), with given entrance and exit velocity profiles. Unfortunately the references for this modified problem are rather scarce.

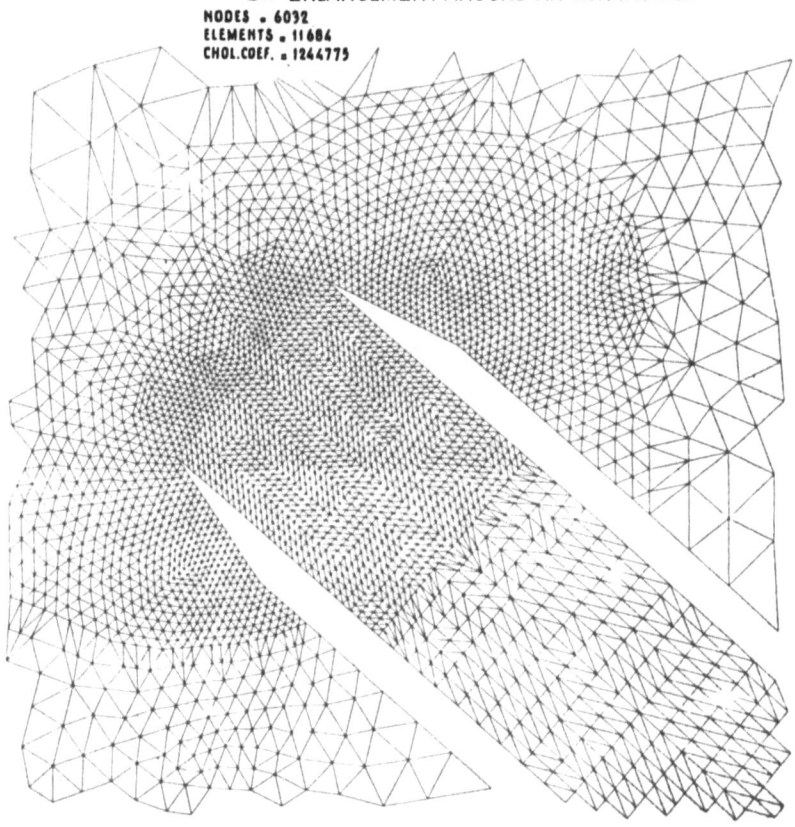

$\widetilde{\mathcal{E}}_h$ ENLARGEMENT AROUND AN AIR INTAKE

NODES = 6032
ELEMENTS = 11684
CHOL.COEF. = 1244775

Fig. 40

## 4.3.2. An Engineering Problem: Unsteady 2-D Flow Around and In an Air-Intake

I am gratefully indebted to J. Periaux from Avions-Marcel-Dassault/Bréguet-Aviation, for giving me this numerical illustration.[88]

The air intake is shown with the meshes on Figures 37, 38, 39 and 40; Figures 37 and 38 show the whole domain while Figures 39 and 40 are enlargements around the air intakes; Figures 37 and 39 show the triangulation used for the conforming $P1$ approximations of the pressure: $15<5$ vertex nodes and 2921 triangles.

On Figures 38 and 40, we see finer mesh used to approximate the velocity field ($4 \times P1$ approximation): 6032 nodes and 11684 elements.

---

[88] The code is due to B. Mantel, from AMD/BA.

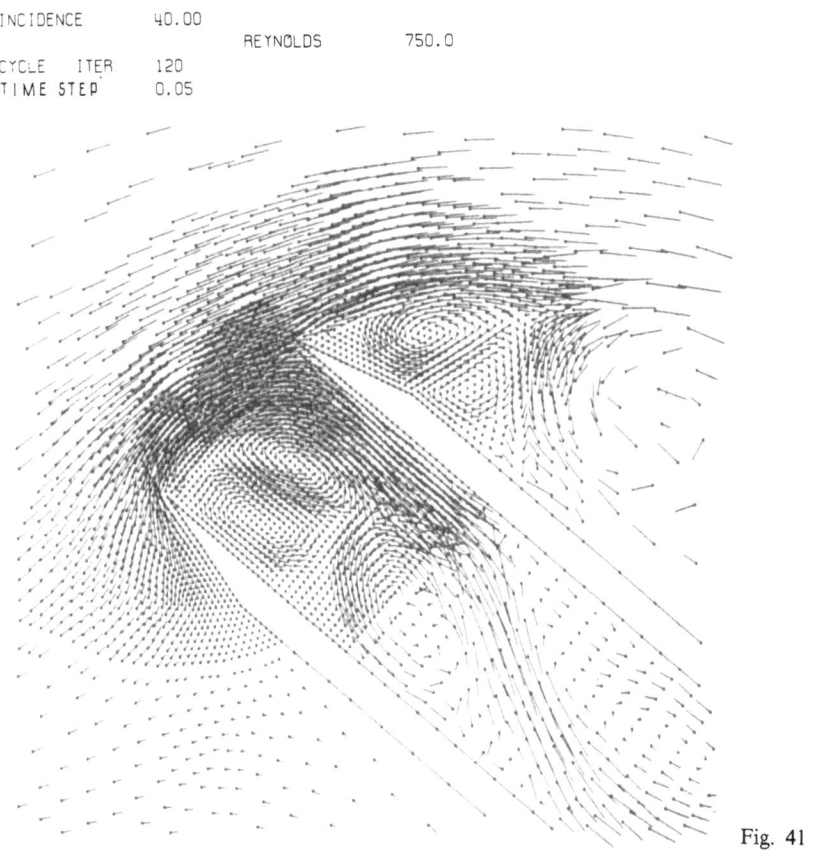

INCIDENCE      40.00
                        REYNOLDS        750.0
CYCLE   ITER   120
TIME STEP      0.05

Fig. 41

The scheme is therefore a variant of "$P2+P1$" approximation, namely the "$4 \times P1 + P1$" element, see §3-3-2. At every time step, the non linear problem is solved via a non linear version of the conjugate gradient method (see chapter 5), so that the non linear problem (for one time step) is reduced to a succession of (about 5 to 10) Stokes problems, which in turn are decomposed into Poisson's equations (§3-3-3).

We might think to use Cholesky factorization to solve these Poisson's equations, but the storage requirements for the factorized matrix would be about $1.2 \times 10^6$ for this problem, which is exceedingly large; therefore Poisson's equations are solved via incomplete Cholesky factorization, accelerated with preconditioned conjugate gradient method. As to the boundary integral equation[89] (135), yielding the pressure on the boundary, it may be solved by Cholesky decomposition (the factorization being done once and for all);

---

[89] With a full $M \times M$ matrix, $M$ = number of boundary points ($M \simeq 120$ in the given example).

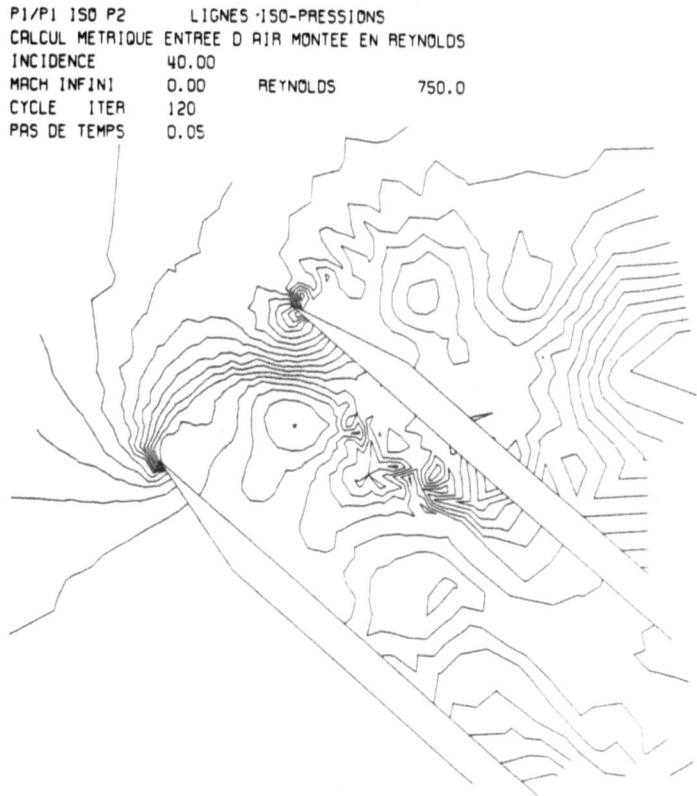

P1/P1 ISO P2      LIGNES ·ISO-PRESSIONS
CALCUL METRIQUE ENTREE D AIR MONTEE EN REYNOLDS
INCIDENCE       40.00
MACH INFINI     0.00      REYNOLDS      750.0
CYCLE   ITER    120
PAS DE TEMPS    0.05

Fig. 42

P1/P1 ISO P2      LIGNES DE COURANT
CALCUL METRIQUE ENTREE D AIR MONTEE EN REYNOLDS
INCIDENCE       40.00
MACH INFINI     0.00      REYNOLDS      750.0
CYCLE   ITER    120
PAS DE TEMPS    0.05

Fig. 43

```
P1/P1 ISO P2      LIGNES ISO-ROTATIONNEL
CALCUL METRIQUE ENTREE D AIR MONTEE EN REYNOLDS
INCIDENCE        40.00
MACH INFINI      0.00      REYNOLDS       750.0
CYCLE   ITER     120
PAS DE TEMPS     0.05
```

Fig. 44

however for a 3-dimensional problem, the use of conjugate gradient is still required for the solution of the boundary equation.

This algorithm with embedded conjugate gradient loops, may seem to have a complex structure; let us emphasize on the strong *stability* properties obtained with this algorithm; moreover the user is not required to give any a priori parameter.[90]

The complexity of the separated flow after 120 time steps ($\delta t = .05$), with large vortex structures moving inside and outside of the inlet, is shown on Figures 41, 42, 43, and 44, via velocity distribution (Figure 41), pressure distribution (Figure 42), stream lines (Figure 43) and vorticity distribution (Figure 44); the incidence angle is 40 degrees, and the Reynolds number 750 (computed on the basis of the inlet width).

This computation required several hours on IBM 370/168.

---

[90] Whose estimation is delicate in other methods.

# 5. Computational Problems and Bookkeeping

This chapter surveys the different problems related to the implementation of finite element methods. We refer to George (1970), Zienckiewicz (1977), Mercier and Pironneau (1977) for more details.

## 5.1. Mesh Generation

Some bookkeeping is necessary to store a finite element mesh: in order to retain the flexibility of finite element methods we can hardly assume a priori properties as regards, for instance, the repartition of nodes (as in finite differences). The minimum information can be stored in two *arrays*:

The first contains the vertex coordinates, which are numbered, say from 1 to $NS$

The second one contains the description of the elements and gives for each element, the numbers of the vertices.

Another array of integers should give information on each vertex, indicating whether this is *interior*, or lies on the *boundary*; alternately we may choose to number the interior vertices from 1 to $NS0$, and the boundary vertices from $NS0 + 1$ to $NS$.

Several families of strategies for generating 2-dimensional triangular meshes for general domains may be considered (cf. Koutchmy et al. (1977) and technical notes of club Modulef):

(i) the obvious one, applicable to special geometries, i.e. generate a *finite difference* mesh and divide each quadrangle into triangles; use may be made of conformal mapping programs developed for the generation of curvilinear finite difference grids.[91] Such a method yields an optimal ordering of the nodes (in view of reducing the bandwidth for the solution of associated linear systems by Cholesky or Gauss factorizations). A related method is considered by Thacker (1977) who perturbs an a priori regular mesh (with equilateral triangles) moving points or deleting triangles, until he obtains concordance with the given physical domain.

---

[91] Of course we may distort the mesh by different simpler (e.g. affine) transformations; e.g. module Quacou in Modulef.

(ii) Define a *coarse mesh*, choose a parameter $n$, and call a subroutine that divides each triangle into $n^2$ new triangles (George (1970), Bourgat (1970); module Retria in Club Modulef). This method is not well suited for the generation of strongly graded meshes, because all the grading is present in the coarse mesh given by the user.

(iii) Fix the *boundary vertices*, and call a subroutine that fills in the domain with triangles, according to Figure 45a to 45f. Among all possible choices the algorithm chooses to construct the triangles that are as close as possible to equilateral triangles: this is the sketch of the method proposed by A. T. George (1970). A simplified version is the module Triaut in Modulef;[92] a related method was developed by J. F. Bourgat (unpublished).

These kinds of methods offer a *great flexibility*, specially when a *graded* mesh is required, but they present several drawbacks.

First the algorithms are comparatively fragile and may be stuck in, principally under the following circumstances:

The user has given boundary vertices with a very strong grading as in Figure 46; in this case the algorithm may begin to generate an exceedingly large number of elements, causing a memory overflow;

The domain is squeezed in the middle as in Figure 47; then the program is no longer able to create triangles in the middle. The remedy is: either to refine the given boundary vertices; or to cut the domain into 2 (or more) pieces, apply the algorithm in each parts and put the parts together as explained in v). Note that such drawbacks are weakened when an interactive graphic display is available.

Moreover the algorithm needs to know where the interior of your physical domain lies; this requires that the boundary vertices make a single connected chain. Finally, no a priori bound (sharper than 20%) can be given for the number of generated triangles.

(iv) Cavendish (1974) proposed to distribute the vertices, in a first step, according to variable densities defined by the user, and in a second step, to connect the generate points to form triangles; Boisvert (1979) recently developed ideas related to the second step.

(v) *Combine* several of the above techniques (or all of them) as follows: partition the physical domain into a number of pieces; apply to each part the appropriate algorithm; put the pieces together (for instance the module Recol in Modulef performs this operation). This should be the most reasonable, and most flexible approach.

(vi) An efficient mesh generating code for quadrangles is given by Jones (1973).

(vii) In 3 dimensions of space, mesh generations into tetrahedra are considered in Zienckiewicz (1977), Marrocco (1978, technical note of Club Modulef), Naves (1975).

---

[92] Coded by P. Borsenberger and modified by A. Perronnet.

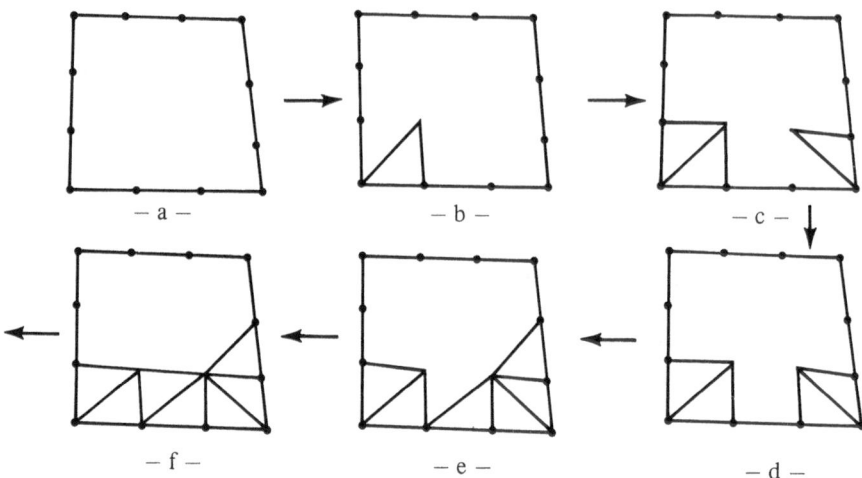

Fig. 45

In general an automatically generated mesh need to be *reordered* so that nodes belonging to one element are indexed by numbers as close as possible; two reasons can be invoked for this reordering operation:

(i) Minimize the bandwidth in view of the use of in core band (or skyline) Cholesky (or LU) solvers);
(ii) Optimize the use of graphic displays.

The most commonly used algorithm was proposed by Cuthill and Mc Kee (1969): the user is required to give an initial set of nodes, defining an "initial front", which are numbered, say from 1 to $NPF$; the algorithm numbers the closest neighbours of the initial front, say from $NPF+1$ to $NPF+NPF'$: thus a "second front" is defined; then the closest neighbours of the second front are numbered, and so forth. The algorithm proceeds until the moving front has swept over the whole domain. See also George (1970), Mercier and Tremolieres (1973) and Modulef technical notes.

Of course, once the mesh is produced, supplementary arrays should (in general) be constructed, e.g. give the numbers of degrees of freedom associated to each element.

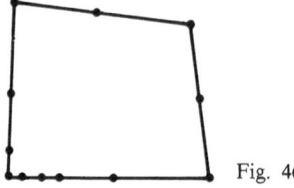

Fig. 46

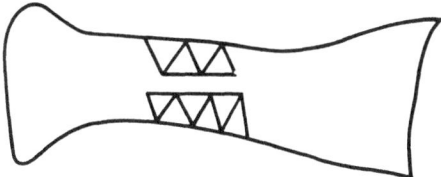

Fig. 47

## 5.2. Solution of the Nonlinear Problems

The time marching schemes for the transient problems will not be discussed because this would fill another book, and this topic is well documented: see e.g. Zienckiewicz (1977), Girault and Raviart (1979), Hughes et al. (1979), and the references of these papers (we will only emphasize on splitting techniques in §5-2-4).

Anyway the use of an implicit or semi implicit schemes requires the solution at each time step of a system similar to the stationary equations, of the following type:

$$\beta \mathbf{u} - \nu \Delta \mathbf{u} + (\mathbf{u} \cdot \nabla) \mathbf{u} = \mathbf{f} - \nabla p$$
$$\operatorname{div} \mathbf{u} = 0 \tag{154}$$

$+$ boundary conditions;

(Of course in the transient case, $\mathbf{f} = $ body force $+ \mathbf{u}^m(\delta t)$ with e.g., $\beta = \delta t^{-1}$, or $\beta = (\frac{3}{2}\delta t)^{-1}$ (in Gear's scheme), $\beta = 0$ for the stationary equations).

$$\beta \mathbf{u} - \nu \Delta \mathbf{u} + (\mathbf{v} \cdot \nabla) \mathbf{u} = \mathbf{f} - \nabla p$$
$$\operatorname{div} \mathbf{u} = 0$$

$+$ boundary conditions; $\tag{155}$

($\mathbf{v}$ given $=$ velocity field at the preceding time step)

The equations (155) yield after finite element discretization, a linear system with non symmetric matrix, see §5-3.

Consider now the various approaches to the solution of the nonlinear equations (154).

### 5.2.1. Successive Approximations (or Linearization) with Under Relaxation

$$\beta \mathbf{w}^m - \nu \Delta \mathbf{w}^m + (\mathbf{u}^{m-1} \cdot \nabla) \mathbf{u}^m = \mathbf{f} - \nabla p^m \cdot$$
$$\operatorname{div} \mathbf{w}^m = 0$$
$$\mathbf{u}^m = \omega \mathbf{w}^m + (1 - \omega) \mathbf{w}^{m-1} \tag{156}$$

($\omega$: relaxation parameter, $0 < \omega < 1$)

This algorithm was analyzed by Crouzeix (1974) and is currently used, particularly when an upwinding scheme is introduced. We get again the same non symmetric system as (155).

### 5.2.2. Newton-Raphson Algorithm

Let us write (154) in matrix form:

$$A \cdot Z + B(Z) \cdot Z = F \tag{157}$$

where $B(Z)$ linearly depends on $Z =$ vector of degrees of freedom. Assume that we know an estimation $Z_{old}$ of $Z$; set:

$$\text{res}(Z_{old}) = A \cdot Z_{old} + B(Z_{old}) \cdot Z_{old} - F = \text{residual vector}$$

We compute the new estimate $Z_{new} = Z_{old} + \delta Z$ by requiring that the new residual $\text{res}(Z_{new})$ should vanish, to the first order in $|\delta Z|$:

$$A \cdot \delta Z + B(\delta Z) \cdot Z_{old} + B(Z_{old}) \cdot \delta Z = - \text{res}(Z_{old}) \tag{158}$$

This is again a system with a non symmetric matrix; after computation of $\delta Z$, replace $Z_{old}$ by $Z_{new} = Z_{old} + \delta Z$, and iterate. When the occurs it is pretty fast, with convergence rates comparable to conjugate gradient methods; unfortunately the domain of convergence (which is very large at low Reynolds numbers) sharpens as the non linearities i.e. Reynolds number, increase.

Furthermore if we use the $LU$ factorization to solve (158), we see that a new matrix is to be assembled and factorized at each time step:

$$K_{new} = A + B(\cdot) + B(Z_{old}).^{93}$$

Alternatives to this situation (Modified Newton-Raphson methods) are surveyed by Matthies and Strang (1979): the tangent matrix $K_0$ is held fixed, at least for several steps; at each step, instead of $K_{new}$ an approximate tangent matrix $K'_{old}$ at the previous step by a low rank (1 or 2) matrix, so that $K'_{new}$ can be updated from $K'_{old}$ through simple rules; only the factorization of $K_0$ is required.

Matthies and Strang tested a method of this type for the analysis of inelastic materials in structural mechanics: they found that the method has better stability properties than the original Newton-Raphson algorithms, without great damage to the rate of convergence.

Further references: Brodlie et al. (1973), Crisfield (1979), Reid (in Himmelblau, 1973). A detailed analysis and comparisons can be found in Bonnet and Meurant (1979).

---

[93] The "tangent" matrix.

### 5.2.3. Conjugate Gradient Method (with Scaling) for Nonlinear Problems

Pironneau (1976), Bristeau et al. (1978, 1979), Glowinski et al. (1979, 1980).

For the sake of simplicity, let us describe the algorithm in the context of Navier Stokes equations (the continuum problem, equations (154)); also in simplification purpose we take homogeneous boundary conditions, so that

$$\mathbf{u} \in V = \left\{ \mathbf{v} \in \left( H_0^1(\Omega) \right)^2; \operatorname{div} \mathbf{v} = 0 \right\}$$

The non linear equations (154) can be written as a least squares formulation in the following way. To any vector valued function $\mathbf{v} \in (H_0^1(\Omega))^2$ we can associate another vector valued function $\mathbf{w} \in (H_0^1(\Omega))^2$ and a function $\Pi$ through the following definition:

$$\beta \mathbf{w} - \nu \Delta \mathbf{w} = \mathbf{f} - \nabla \Pi - (\mathbf{v} \cdot \nabla) \mathbf{v}$$

$$\operatorname{div} \mathbf{w} = 0 \tag{159}$$

so that $\mathbf{w} = \mathbf{w}(\mathbf{v})$, $\Pi = \Pi(\mathbf{v})$ (note that (159) is a Stokes problem in $\mathbf{w}$).

If $\mathbf{w} = \mathbf{v}$, then the original problem is solved, with $\mathbf{u} = \mathbf{v} = \mathbf{w}$; therefore we try to minimize the difference $\mathbf{v} - \mathbf{w}(\mathbf{v})$, in a norm adapted to the problem. More precisely we set the following *minimization* problem:

$$\operatorname{Min} J(\mathbf{v}) \tag{160}$$

$$\mathbf{v} \in V$$

where the cost functional is defined by:

$$J(\mathbf{v}) = \int_\Omega \left[ \beta |\mathbf{v} - \mathbf{w}(\mathbf{v})|^2 + \nu |\operatorname{grad}(\mathbf{v} - \mathbf{w}(\mathbf{v}))|^2 \right] dx$$

For convenience we shall note $[\cdot, \cdot]$ the associated scalar product:

$$[\mathbf{u}, \mathbf{v}] = \int_\Omega (\beta \mathbf{u} \cdot \mathbf{v} + \nu \operatorname{grad} \mathbf{u} \cdot \operatorname{grad} \mathbf{v}) \, dx \tag{161}$$

(so that $J(\mathbf{v}) = [\mathbf{v} - \mathbf{w}(\mathbf{v}), \mathbf{v} - \mathbf{w}(\mathbf{v})]$)

The problem is solved via a version of the conjugate gradient method (Polak (1971)), in a metric defined precisely by this scalar product:

**Step 0:** (*initialization*) Choose an initial velocity field $\mathbf{u}_0$ (the solution at previous time step in the transient problem or, say, the Stokes solution in the case of the stationary problem); then compute the *gradient* of the functional (solve a Stokes problem)

$$[\mathbf{g}_0, \mathbf{z}] = \langle J'(\mathbf{u}_0), \mathbf{z} \rangle, \qquad \forall \mathbf{z} \in V$$

$$\mathbf{g}_0 \in V$$

set $\zeta = \mathbf{g}_0$; (see the computation of the gradient further below).

**Step 1:** *Descent*: minimize $J(\cdot)$ along the line defined by the direction $-\mathbf{g}_0$, i.e., find $\lambda>0$ such that:

$$J(\mathbf{u}_0-\lambda\boldsymbol{\zeta}_0)\leq J(\mathbf{u}_0-\mu\boldsymbol{\zeta}_0), \qquad \forall\mu$$

(This line search can be achieved by dichotomy or Fibonacci methods; in practice it involves 3 or 4 evaluations of the cost functional).

Then, set:

$$\mathbf{u}_n=\mathbf{u}_0-\lambda\boldsymbol{\zeta}_0;$$

test the convergence

**Step 2:** *Construction of the new descent direction*

$$\mathbf{g}_n\in V(\text{new gradient})$$

$$[\mathbf{g}_n,\mathbf{z}]=\langle J'(\mathbf{u}_n),\mathbf{z}\rangle, \qquad \forall\mathbf{z}\in V.$$

$$\gamma=\frac{[\mathbf{g}_n,\mathbf{g}_n-\mathbf{g}_0]}{[\mathbf{g}_0,\mathbf{g}_0]}$$

$$\boldsymbol{\zeta}_n=\mathbf{g}_n+\gamma\boldsymbol{\zeta}_0$$

$$\boldsymbol{\zeta}_0=\boldsymbol{\zeta}_n,\mathbf{u}_0=\mathbf{u}_n, \text{ go to step } 1. \qquad \square$$

At each step we have to compute the gradient of the cost functional: this is defined through the first variation of $J(\mathbf{v})$; by definition:

$$J(\mathbf{v}=\delta\mathbf{v})-J(\mathbf{v})=\langle J(\mathbf{v}),\delta\mathbf{v}\rangle+0(|\delta\mathbf{v}|^2)$$

After some manipulations (see the details in Bristeau et al. (1978, 1979)), we find $(\mathbf{w}=\mathbf{w}(\mathbf{v}))$:

$$\langle \dot{J}'(\mathbf{v}),\delta\mathbf{v}\rangle=[\mathbf{v}-\mathbf{w},\delta\mathbf{v}]+\int_\Omega[(\mathbf{v}\cdot\nabla)\delta\mathbf{v}]\cdot(\mathbf{v}-\mathbf{w})\,dx$$

$$+\int_\Omega[(\delta\dot{\mathbf{v}}\cdot\nabla)\mathbf{v}]\cdot(\mathbf{v}-\mathbf{w})\,dx$$

(As we have said the computation of the gradient $\mathbf{g}_n$ involves the solution of a Stokes problem.)

From the numerical experiments it appears that the convergence is fast;[94] for instance to compute the stationary flow in the square cavity at $R=500$, with a regular $20\times20$ mesh ("$4\times P1+P1$" element), the convergence was obtained in 50 iterations.

Nevertheless, as the Reynolds number increases for a fixed mesh the convergence slows down: indeed if the mesh is to coarse for the representation of the flow, it is practically impossible to bring the cost functional to zero (or to an acceptably small value), in which case $\mathbf{u}_n-\mathbf{w}(\mathbf{u}_n)\neq0$, and the problem is not solved (this is the case for the computations in the square

---

[94] The method is both faster than successive approximation and safer than Newton-Raphson's.

cavity at $R = 10^4$ given in Bristeau et al. (1979)). However even when this non convergence occurs, the algorithm remains *stable* (since the cost functional can only decrease); furthermore even a wrong computation gives indications on the region where a refinement is required (for a resubmit of the job) if it was not guessed a priori: indeed the contributions to the non convergence of the cost functional can be localized.

Further references: Daniel (1971), Concus, Golub and O'Leary (1977).

The method was originally developed for the computation of transonic flows (cf. e.g. Bristeau (1978)).

### 5.2.4. A Splitting Technique for the Transient Problem

For transient Navier Stokes equations, Benque et al. (1979, 1980) propose a *splitting* technique, so as to separate in the equations the convective and diffusive terms: ($m$ = time counter)

(advection) $\quad \dfrac{\mathbf{u}^* - \mathbf{u}^m}{\delta t} + (\mathbf{u}^m \cdot \nabla)\mathbf{u}^* = \mathbf{f}$

$$\frac{\mathbf{u}^{m+1} - \mathbf{u}^*}{\delta t} - \nu \Delta \mathbf{u}^{m+1} = -\nabla p^{m+1}$$

(Stokes)
$$\operatorname{div} \mathbf{u}^{m+1} = 0,$$
$$+ \text{boundary conditions.}$$

The advection equation is solved via the method of characteristics, §2-6; the solution of Stokes equations can make use of any efficient Stokes solver, e.g. the decomposition of Stokes problem, §3-3-3 (but other choices could be made).

The method is *unconditionally stable*.

## 5.3. Iterative and Direct Solvers of Linear Equations

As we have seen the application of finite element methods finally amounts to the solution of (usually) several successive systems of linear equations, with a spare, sometimes symmetric and positive definite, matrix. Several families of classical solvers are known:

### 5.3.1. Successive Over Relaxation

Varga (1962), Young (1971)

This method benefits from some popularity because it requires a minimum programming effort, and only the non zero elements of the matrix need to be stored. Unfortunately it suffers from severe drawbacks:

The *convergence rate* is very sensitive to the choice of the acceleration parameter, whose value cannot be estimated, but for an empirical search (in contrast to the matrix arising from a finite difference method);

Even if the *optimal* value of the acceleration parameter is known, the tests performed by D. Kershaw (1979) show that the conjugate gradient method with scaling (incomplete Cholesky, see §5-3-3 below) was about 30 times faster for a typical problem;

Anyway for a non linear problem such as Navier Stokes equations (with an implicit time marching scheme), the matrix, and therefore the acceleration parameter $\omega$ depends on the iteration; so that even if $\omega$ were optimal at the beginning, there is practically no chance that it remains so throughout the computation.

(For extensions of Sor, see Axelsson (1976).)


### 5.3.2. Cholesky Factorizations

Consider the case where the system to be solved is of the form:

$A \cdot x = f$

with $A$ symmetric and positive definite. Then $A$ can be factored as:

$A = L \cdot L^T$

where $L$ is lower triangular and $L^T =$ transpose of $L$. Then the solution is found in two steps:

i) $L \cdot y = f$
ii) $L^T \cdot x = y$

(the solution of a linear system with a lower or upper triangular matrix is straightforward since each equation can be solved in turn, with the resulting values substituted into the remaining unsolved equations).

A variant is the factorization:

$A = L \cdot D \cdot L^T,$

with $D$ a diagonal matrix and the diagonal elements of $L$ set to 1: this form avoids to call for square root extractions during the factorization process. Now advantage should be taken of the sparsity of $A$: with finite element discretizations, usually $a_{ij} = 0$ when $|i-j| > b$ (= the bandwidth of the matrix). If the nodes are correctly ordered[95], then $b$ is significantly less than the number of equations, although $b$ may become quite large in a 3-dimensional problem. It can be shown that $L$ inherits the band structure of $A$ (Martin and Wilkinson (1965, 1967)): $L$ does not have non zero elements *outside* the bandwidth, but that it suffers from *fill-in inside* the bandwidth; that is to say, if $a_{i,i+k} \neq 0$, even if the other elements vanish in row $i$ between columns $i+1$

---

[95] Note the importance of reordering the mesh after its generation.

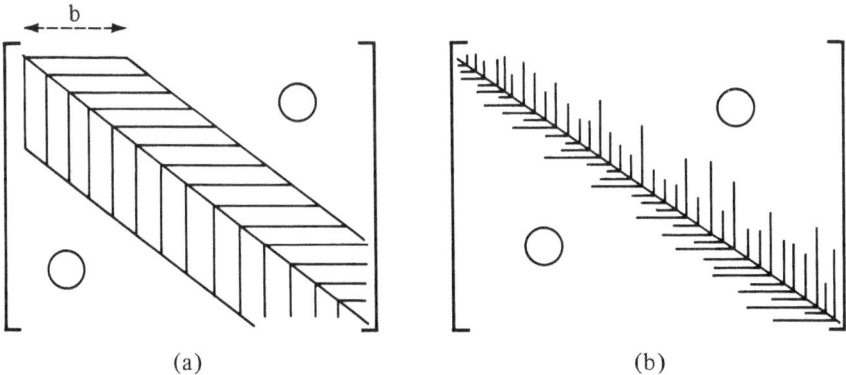

Fig. 48 (a) A band structured matrix[96]; (b) A profile structured matrix[96]

and $i+k-1$, nevertheless a priori $l_{ij}\neq0$ in the corresponding positions $(i\leq j\leq i+k)$.

Storage requirements can be reduced (usually by $\sim30\%$), and also the computation time, by the use of a profile (or skyline) storage scheme (Jennings (1966)): consider for each column $j$ the topmost nonzero element of $A$ (above the diagonal); this lies in row $r(j)$:

$$a_{ij}=0 \quad \text{for } i<r(j).$$

Then $L$ and $L^T$ inherit the profile structure of $A$, i.e. $u_{ij}$ (coefficient of $L^T$)$=0$ for $i<r(j)$ but $u_{ij}$ is—a priori—nonzero if $r(j)\leq i\leq j$. The fill-in of the profile for a typical matrix is illustrated on Figure 49.

Note the dramatic improvement brought in by profile storage—over band storage—, in the problem of 2 dimensional flow around an obstacle (or several obstacles) using any $\psi$-$\omega$ formulation: the stream function is held constant, with an unknown value on each obstacle; there exist for each obstacle a basis function whose support surrounds the whole obstacle (it is "less" local than for the ordinary basis functions); for the corresponding column $j$, there is no lower bound on $r(j)$ which might be, say equal to 1, so that the bandwidth is pretty large. On the other hand a skyline scheme will involve only the storage of one supplementary column per obstacle. (Similar storage problems are met in case of periodic boundary conditions.) However even with skyline storage schemes the storage requirements for the factorized $L$ is significantly larger than for the non zero elements of the original matrix $A$.

In large[97] engineering problems the fill in may cause memory overflow and this prompted the development: of out-of core methods on one hand (use of

---

[96] Storage in core locations are required for the coefficients situated in the hatched regions of this figure.

[97] 3-D problems *are* large if reasonable accuracy is wanted.

# CHOLESKY FACTORIZATION

$$A_h = L_h L_h^t$$

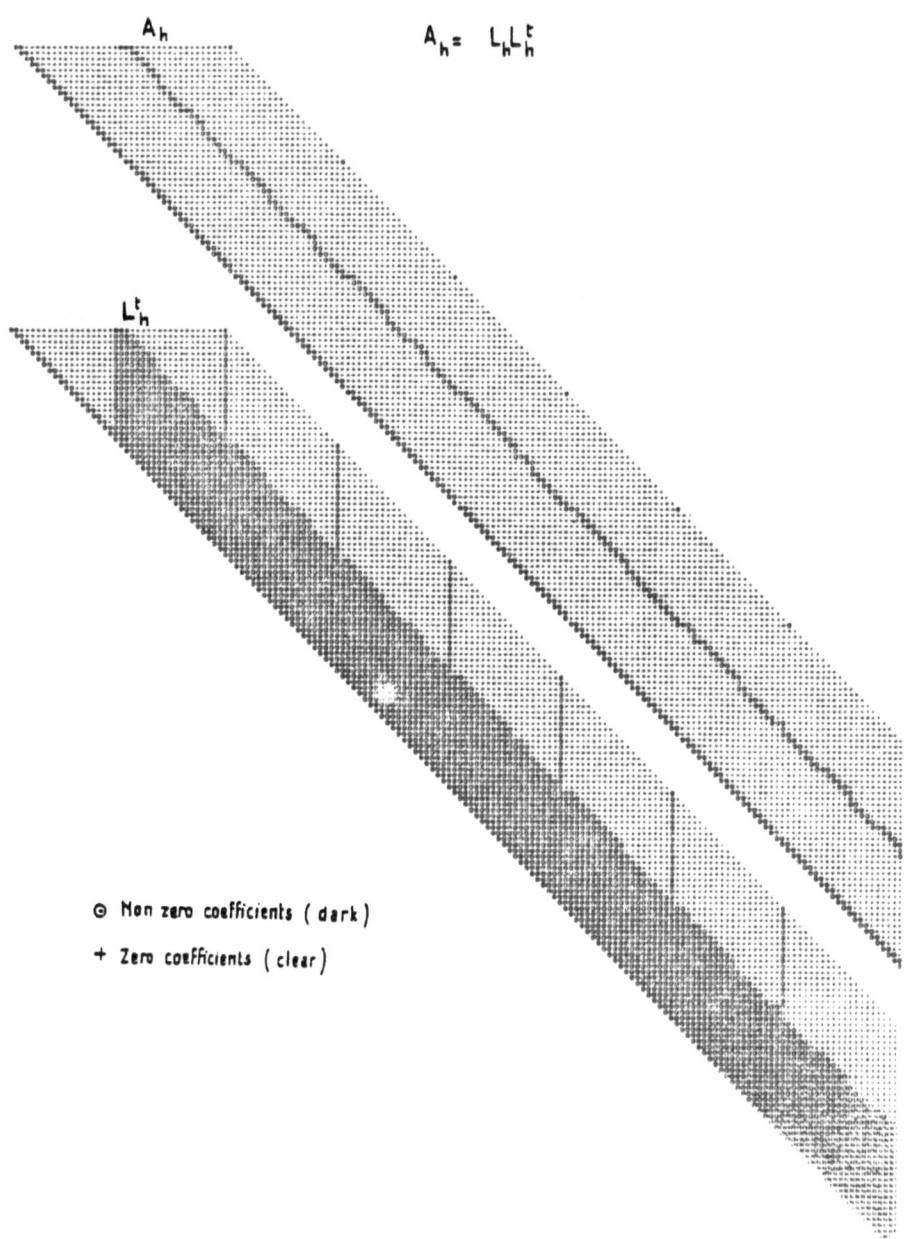

$\odot$ Non zero coefficients ( dark )

$+$ Zero coefficients ( clear )

Fig. 49

secondary storage) (§5-3-3) recently of incomplete Cholesky factorizations on the other hand (§5-3-4).

Of course when $A$ is non symmetric, $A$ is factorized as:

$$A = L \cdot U$$

with $L =$ lower triangular and $U =$ upper triangular matrices; $L$ and $U$ inherit the band or profile structure of $A$ as in the symmetric case (this factorization is equivalent to Gauss elimination).

Further references: George (1970), Duff (1979), Tewarson (1970, 1973), Bunch and Rose (1976), Reid (1971), Rose and Willoughby (1972), Himmelblau (1973), and Modulef technical notes.

The stability of $LU$ factorization (together with possible pivotal recordings preserving the sparsity) is analyzed in Tewarson (1970), Curtis and Reid (1971), Erisman and Reid (1974).

### 5.3.3. Out of Core Factorizations

In a first stage, the matrix can be assembled, *partitioned into blocks*, with only one block present at a time in central core memory, the rest lying on secondary storage (tape or disk); in a second stage the blocks are read for factorization: Cantin (1971), Wilson et al. (1974), Cheung and Khatua (1976), Mondkar and Powell (1974), Van Ingelandt (1979), Hasbany and Engelman (1979).

Such methods offer practically unlimited possibilities as to the size of systems to be handled; however they make use of *random access* (non sequential) read/write instructions, so that the waiting time (during which the job is inactive and waiting for the end of input/output operation) can become important.

Alternatively the family of methods issued from the ideas of B. M. Irons (1970) or Aufaure and Benjamin (1970), namely the *frontal* method and the related ones, use *sequential access* disk or tape files.

The method lies on a careful examination of the Gauss elimination process: you can in the same pass, begin the assembling of the matrix, start the Gauss elimination process on lines fully assembled, and send these lines to the auxiliary storage.

For a brief account of the method, consider the $Q1$-approximation of, say a Poisson's equation, with the mesh shown on Figure 50. Assuming Neumann boundary conditions, the discrete equations take the form:

$$\sum_{j=1}^{N} a_{ij} x_j = f_i, \qquad 1 \le i \le N (N = 16)$$

The coefficients of the linear system take the form:

$$a_{ij} = \sum_{K/i,j} a_{ij}^K, \qquad f_i = \sum_{T/i} f_i^T$$

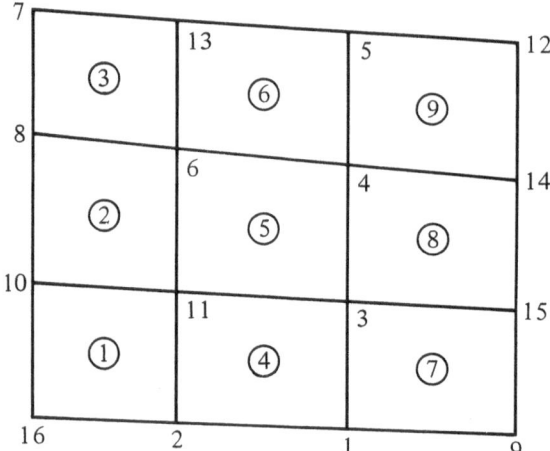

Fig. 50

where the summation is extended to those elements adjacent to nodes $i$ and $j$ (otherwise $a_{ij}^K$ is zero).

Now let us perform the assembling of equations, one element after another, in the order indicated on the Figure. We see that already after the assembling of the element 1, the equation corresponding to node 16 is fully summed, since node 16 is absent of subsequent elements. Now we do not know yet the rest of the matrix since we only know the contributions of element 1, nevertheless the Gaussian elimination can be started; indeed this involves the following operations on the full matrix:

$$a_{ij} := a_{ij} + \tilde{a}_{ij}, \qquad (i \text{ and } j \neq 16)$$
$$\tilde{a}_{ij} := -a_{i,16} a_{16,j} / a_{16,16}$$

(and similar operations on the right hand side).

Thus: either $a_{i,16} = 0$ or $a_{16,j} = 0$ (node $i$ or $j$ is not neighbour to node 16), in which case we do not need to bother for ignoring at this stage the value of $a_{ij}$; or we can compute $\tilde{a}_{ij}$ (nodes $i$ and $j$ are neighbour of node 16) and store the result in a core memory location (the contributions to $a_{ij}$ from other elements will be summed in time). After that we no longer need the row number 16 and we can write it on auxiliary storage, thus some core memory is freed (this allocation/desallocation process involves some bookkeeping). We must keep in core memory locations for the lines 2, 11, 10, which are not fully summed. Next we pass on the element 2: after assembling the Gauss elimination can be performed with line 10 as pivot row; afterwards the row number 10 can be written on auxiliary storage; we keep in core location corresponding to rows 2, 11, 6, 8 which are not fully summed, and so forth.

The algorithm proceeds on, until all elements have been considered, and Gaussian elimination performed on all rows: then the file is to be read backwards in order to obtain the solution by packsubstitution.

**Remarks.**

(i) The core memory requirement is dictated by the frontwidth, i.e. by the number of nodes that may be active at a time.

(ii) The elements need to be well ordered, so as to minimize the frontwidth; special variants of Cuthill and Mc Kee's algorithm are available for this task (e.g. module Renel in Modulef).

(iii) In contrast to skyline or band methods, the fill-in is of no importance.

(iv) If enough core memory is available, we can wait and assemble a pack of elements before performing Gauss eliminations: this will allow for more possibilities of pivoting strategies and furthermore reduces the number of read/write operations on the file.

(v) Hood (1976) extended Irons' method to $Lu$ factorization of a non symmetric matrix, with pivoting.

Further references and related methods: Aussems (1972), Joly (1979), Cooke (1978), Vendham et al. (1974), Mondkar and Powell (1974), Bossavit and Fremond (1976).

In spite of its attractive properties the cost of the frontal method may be prohibitive for very large problems for two kinds of reasons:

(i) The frontwidth may be so large that even the front matrix cannot be stored in the core memory (this is specially dramatic when "mixed" methods are to be used for an engineering problem);

(ii) Waiting time due to read/write operations.

### 5.3.4. Preconditioned Conjugate Gradient

The conjugate gradient method was proposed by Hestenes and Stiefel (1952) for the solution of a linear system $Ax=b$, with $A$ a symmetric, positive definite, $N \times N$ matrix: iterates $x^m$, $m=0,1,2,\ldots$, are constructed so that the residuals $r^m = Ax^m - b$ are mutually orthogonal. Thus $r^N = 0$: the solution is theoretically obtained in $N$ iterations; yet this result is handly useful since in practical problems $N$ may range from $0(10^2)$, to $0(10^5)$ (in three dimensional problems). The method regained some favour when it was observed that, as an iterative method, the conjugate gradient method converges very fast provided the *condition number*[98] is close to *one* (Reid (1971), Golub, in Glowinski and Lions (1975).

Thus the conjugate gradient method should be applied to a modified matrix $A = B^{-1} A$ with eigenvalues as close to 1 as possible and where matrix $B$ is easily invertible (intuitively, the closer $C$ is to $A$, the faster the convergence).

In order to construct $C$ one can do the *incomplete Cholesky factorization*:

$$B = LDL^T, \qquad A = B + R;$$

---

[98] I.e. the ratio of the greatest to smallest eigenvalue of $A$.

a given sparsity pattern is forced on $L$; that is a subset of entries $(i, j) \in P$ is chosen.

One proceeds through the ordinary Cholesky algorithm and whenever a coefficient $l_{ij}$ turns up with $(i, j) \in P$, this $l_{ij}$ is forced to zero. The simplest choice is to take the same sparsity pattern for $L$ as for $A$:[99]

$$P = \{(i, j): a_{ij} = 0\}, \quad \text{i.e. } a_{ij} = 0 \Rightarrow l_{ij} = 0$$

Meijerink and Van Der Vorst (1977) proved that if $A$ satisfies $a_{ij} \leq 0$ for $i \neq j$, ($M$-matrix) this choice yields a positive definite approximate matrix $B = LDL^T$ (in particular $\alpha_{ii} > 0$). Unfortunately the $M$-matrix property is seldom satisfied (except when P1-triangle approximation is used for Poisson's equation). When $A$ is not an $M$-matrix simple counter examples show that the algorithm may break down yielding $\alpha_{ii} \leq 0$ at some stage; Kershaw (1978) suggests, when such an event occurs, to force $\alpha_{ii}$ to some small positive value and reports successful results.

Alternatively, if many systems are required to be solved with the same matrix $A$[100] it is worthwhile to spend some time to construct an improved factorization intermediate between the Meijerink-Van Der Vorst factorization, and the complete Cholesky factorization. Therefore Periaux (1979) constructs the complete Cholesky factorization in a first step:

$$A = LL^T$$

(possibly requiring the use of auxilliary storage); then, given a positive constant $c$, define $\tilde{L}_c = [\tilde{l}_{ij}]$ by:

$$\tilde{l}_{ij} = 0 \text{ if } a_{ij} = 0 \quad \text{and} \quad |l_{ij}| \leq c \min_{i, j} \{l_{ii}, l_{jj}\}$$

$$\text{otherwise } \tilde{l}_{ij} = l_{ij}$$

$$\left(L = [l_{ij}]\right)$$

Thus $\tilde{L}_c$ is always guaranteed to be non singular.

Other rules are possible to form $\tilde{L}_c$: see e.g. Bristeau et al. (1979), Glowinski, Mantel et al. (1979). Practically only $\sim 5\%$–$25\%$ of the storage requirement for $L$ (complete Cholesky factorization) can be retained to form $\tilde{L}_c$, with a resulting number of conjugate gradient iterations of $15 \sim 20$.

Now the conjugate gradient algorithm with metric defined by $B$ runs as follows:

Step 0: find an initial guess $x^0$

$$Z^0 = B^{-1}(b - Ax^0)$$

$$P^0 = Z^0$$

---

[99] So that the requirements in storage and CPU time are considerably reduced, compared to the formation of complete Cholesky factorization.

[100] As it is the case when Stokes/Navier-Stokes equations are decomposed into a cascade of Poisson's equations. §3-3-3, §4-3-2, §5-2-3.

Step 1: $\rho = (Z^{0T}BZ^0)/(p^{0T}Ap^0)$
Step 2: $x^1 = x^0 + \rho p^0$
Step 3: $Z^1 = B^{-1}(b - Ax^1)$
Step 4: $\gamma = (Z^{1T}BZ^1)/(Z^{0T}BZ^0)$
Step 5: $p^1 = Z^1 + \gamma p^0$
Step 6: $p^0 := p^1; Z^0 := Z^1;$
Step 7: check for convergence: go to step 1 or end.

Further references: Concus and Golub (in Glowinski and Lions, 1975).

Systematic comparisons and analysis of such methods is given by Bonnet and Meurant (1979). Kershaw (1978) proposed an extension to $LU$ factorization in the non symmetric case.

For the application to symmetric indefinite systems see Axelsson and Gustafsson (1979).

From the mentioned papers, it appears that preconditioned conjugate gradient is the fastest of in core methods.

# Appendix 1. The Patch Test of the *P1* Nonconforming Triangle: Sketchy Proof of Convergence

For the sake of simplicity, we consider Poisson's equation:

$$-\Delta u = f \quad \text{in } \Omega$$
$$u|_\Gamma = 0$$

with the variational formulation:

$$\int_\Omega \text{grad } u \cdot \text{grad } v \, dx = \int_\Omega fv \, dx, \qquad \forall \, v \in H^1_0(\Omega)$$

Also we assume that $\Omega$ is a polygon, so that it is discretized by triangles without any error.

We set:

$V_h = v_h$ such that
        $-v_h$ is piecewise polynomial degree $\leq 1$;
        $-v_h$ is continuous at the mid side nodes;
        $-v_h = 0$ at the boundary nodes

$$a_h(u_h, v_h) = \sum_T \int_T \text{grad } u_h \cdot \text{grad } v_h \, dx$$

$$\|v\|_h = (a_h(v, v))^{1/2} \quad \text{(discrete energy norm)}.$$

Then the discrete problem is:

Find $u_h \in V_h$ such that: $a_h(u_h, v_h) = \int_\Omega fv_h \, dx \; (\equiv (f, v_h))$

We want to estimate the error $\|u - u_h\|_h$. We follow Ciarlet (1975).
By the usual triangular inequality, for any $v_h \in V_h$:

$$\|u - u_h\|_h \leq \|u - v_h\|_h + \|u_h - v_h\|_h \tag{A1}$$

Now, we can write from the definition of $\|\cdot\|_h$

$$\|u_h - v_h\|_h^2 = a_h(u_h - v_h, u_h - v_h) = a_h(u_h, u_h - v_h) - a_h(v_h, u_h - v_h)$$

Since $u_h - v_h \in V_h$,

$$a_h(u_h, u_h - v_h) = (f, u_h - v_h) = (-\Delta u, u_h - v_h)$$

Thus:

$$\|u_h - v_h\|_h^2 = (-\Delta u, u_h - v_h) - u_h(u, u_h - v_h) + a_h(u - v_h, u_h - v_h)$$

If we set:

$$E(u, w_h) = (-\Delta u, u_h - v_h) - a_h(u, u_h - v_h)$$

then,

$$\|u_h - v_h\|_h^2 = E(u, u_h - v_h) + a_h(u - v_h, u_h - v_h)$$

By the Schwartz' inequality:

$$\|u_h - v_h\|_h \leq \|u - v_h\|_h + \frac{|E(u, u_h - v_h)|}{\|u_h - v_h\|_h}$$

$$\|u_h - v_h\|_h \leq \|u - v_h\|_h + \operatorname*{Sup}_{w_h \in V_h} \frac{|E(u, w_h)|}{\|w_h\|_h} \tag{A2}$$

We note that this inequality is true for any $v_h$.

Combining (A2) with (A1) we get the following result (Nitsche (1974), Lascaux and Lesaint (1975)):

$$\boxed{\|u - u_h\|_h \leq \left( \operatorname*{Inf}_{v_h \in V_h} 2\|u - v_h\|_h \right) + \left( \operatorname*{Sup}_{w_h \in V_h} \frac{|E_h(u, w_h)|}{\|w_h\|_h} \right)} \tag{A3}$$

The first term in the right hand side can be shown to be $O(h)$ from classical interpolation results; the second term measures the "*non conformity*" of the method: indeed, using Green's formula on each triangle, we get:

$$E_h(u, w_h) = \sum_T \int_{\partial T} -\frac{\partial u}{\partial u_T} w_h \, ds \tag{A4}$$

($\partial T$ = boundary of $T$, $n^T$ = *exterior* unit normal vector to $\partial T$, $\partial u/\partial n_T = u_{,i} n_i^T$); thus $E_h$ would vanish if $V_h \subset V$.

We refer to Ciarlet (1975) for the end of the proof: $\|u - u_h\|_h = O(h)$. We just point out that (A3) can be viewed as a variational form of Irons' patch test: indeed, if $u$ happens to be a polynomial of degree 1 on the whole domain, then from (A3) and (A4), $u_h - u = 0$.

**Remark.** As Stummel (1979) has pointed out, Irons' patch test is not a sufficient condition for convergence; Lascaux and Lesaint's proof implies that the patch test is passed but the converse is not true (a counter example can be found in Stummel's paper).

# Appendix 2. Numerical Illustration

(J. F. Bourgat, unpublished).

The following equation is considered:

$$-\operatorname{div} a(x)\operatorname{grad} u_\varepsilon = f \quad \text{in } \Omega$$

$$u_\varepsilon|_\Gamma = 0$$

$\Omega$ is a square $)0, 100(\times)0, 100($ $a(x)=1$, except in a narrow band of width 2 and parabolic shape, where $a(x)=1/\varepsilon$ (see the domain and triangulation on Figures 9c, 9d; 9e is a "zoom" show the mesh refinement in the narrow band).

$$f(x) = 52.5 \times x$$

The solution process used $P1$ conforming approximation, and SOR solver (about 20 iterations); for $\varepsilon = 10^{-4}$, the variation of the computed solution is shown of Figure 9e: as expected the solution exhibits large gradients within the narrow central band. This solution coincides, to machine precision with the solution of asymptotic methods. We emphasize on the fact that the regions of strong gradient can be a priori exactly located; this is *not* the case in a lot of practical problems.

However, this example shows that triangles with pretty acute angles can yield accurate precision. Of course this process cannot be used for exceedingly small $\varepsilon$ (e.g. $\varepsilon = 10^{-20}$ when single precision (8 bytes words) is used), unless ill conditioning of the matrix generates show convergence of SOR and/or unacceptable round off errors: in this case only the asymptotic method (or the upwinding techniques of Chapter 2) can yield a solution.

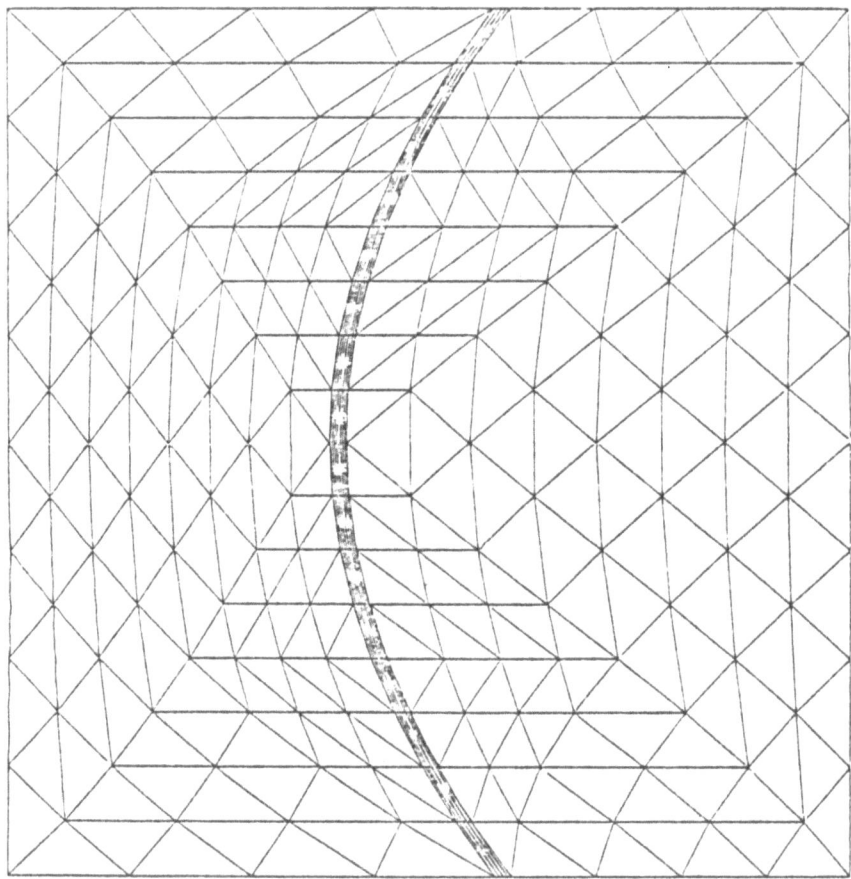

Fig. 9c 544 triangles

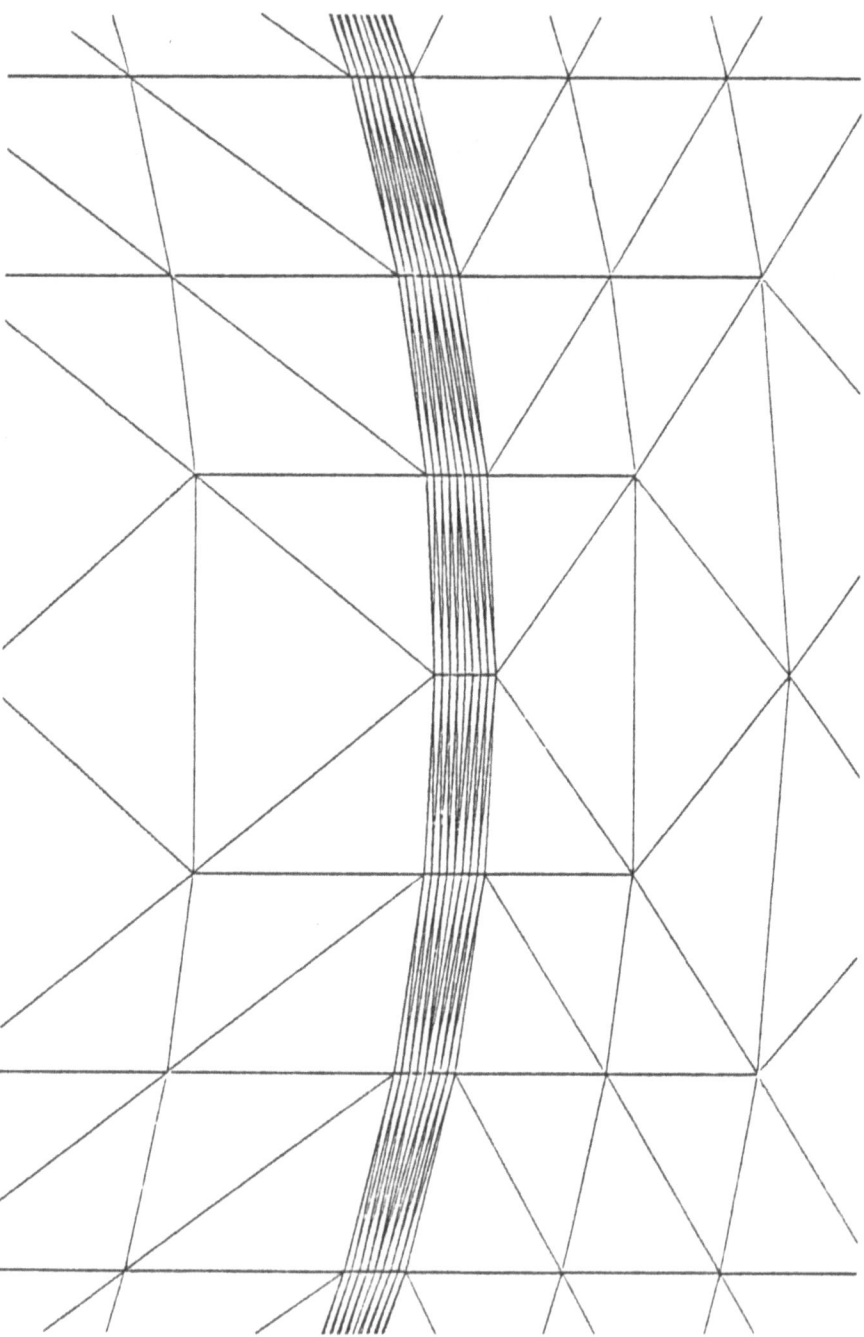

Fig. 9d  zoom from Figure 9c

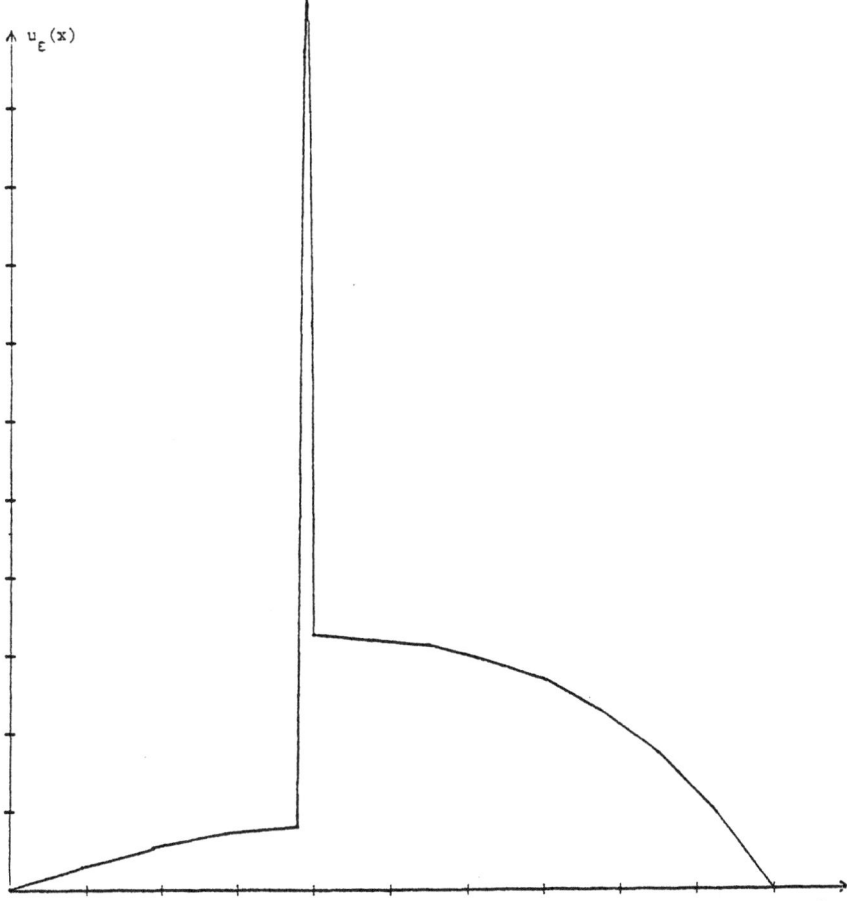

Fig. 9e variation of $u_\varepsilon(x)$ along $y=50$, $\varepsilon=10^{-4}$

# Appendix 3. The Zero Divergence Basis for 2-D $P_1$ Nonconforming Elements

(cf. pp. 96–100)

We prove thereafter that the set of functions $\{w_m, w_s\}$ defined in §3.2.1 makes a basis of:

$$V_h = \{v_h: \; -v_h \text{ is piecewise polynomial of degree } \leq 1;$$
$$-v_h \text{ is continuous at every mid side node;}$$
$$-\operatorname{div} v_h = 0 \text{ on each triangle}\}$$

At each mid side node, we can decompose each $v_h \in V_h$ into its normal and tangent (with respect to the side direction) components:

$$V_h = V_h^1 \oplus V_h^2,$$

where the functions of $V_h^1$ (resp. $V_h^2$) are normal (resp. parallel to each side at the mid point. The set $\{w_m\}$ defined page 77, is obviously a basis of $V_h^2$; we will prove that, if $s_0$ is any fixed vertex, $\{w_s, s \neq s_0\}$ is a basis of $V_h^1$, following the technique of Hecht (1980).

First we require some notations.
Let $S$ be the set of vertices in the mesh;

$$N = |S| = \text{number of vertices.}$$

To any couple $\{a, b\}$ of vertices we associate a vector $\gamma^{a,b} \equiv \gamma^{b,a} \in \mathbb{R}^N$:

$$\gamma_s^{a,b} = 0 \text{ if the vertex } s \in S \text{ is different of } a \text{ and } b;$$

$$\gamma_a^{a,b} = +1$$
$$\qquad\qquad \text{arbitrary choice between } a \text{ and } b$$
$$\gamma_b^{a,b} = -1$$

Then we define the *flux* of $v_h \in J_h^1$ going through between $a, b$ as follows: Consider any *path* from $a$ to $b$, made of triangle sides and let $\mathbf{n}$ be the unit normal vector pointing towards the left of the path if $\gamma_a^{a,b} = +1$, to the right otherwise (see figure).

Then we set:

$$\psi(a, b; v_h) = \int_a^b v_h \cdot \mathbf{n}\, ds \qquad (= -\psi(b, a; v_h))$$

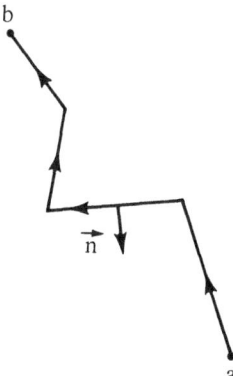

**Remark.** Thanks to the incompressibility of $v_h$, $\psi(a, b; v_h)$ is independent of the path chosen between $a, b$.

**Lemma 1.** $\psi(a, b; w_s) = \gamma_s$,     $\forall s \in S$.

PROOF. Elementary from the definitions.

**Lemma 2.** *Set $s_0$ be some fixed vertex in $S$.*

$$\forall\, v_h \in V_h^1, v_h = \sum_{s \neq s_0} \gamma_s^{s_0, s} \psi(s_0, s; v_h) w_s$$

PROOF. For any $v_h$ in $V_h^1$, set:

$$w_h = \sum_{s \neq s_0} \gamma_s^{s_0, s} \psi(s_0, s; v_h) w_s$$

Consider any *side* between 2 adjacent vertices $a, v$; with midpoint $m$. Because of the linearity of $\psi$ with respect to $v_h$, and of lemma 1:

$$\psi(a, b; w_h) = \sum_{s \neq s_0} \gamma_s^{s_0 s} \psi(s_0, s; v_h) \gamma_s^{ab}$$

Since $|\gamma_s^{ab}| = 1$ if $s = a$ or $s = b$, and $\gamma_s^{ab} = 0$ otherwise, it is readily checked that:

$$\psi(a, b; w_h) = \psi(a, b; v_h)$$

Therefore if $m$ is the mid point of $(ab)$ and $\mathbf{n}$ the normal direction to $ab$:

$$w_h(m) \cdot \mathbf{n} = v_h(m) \cdot \mathbf{n}$$

Since $v_h$ and $w_h \in V_h^1$, this implies the equality of $w_h$ and $v_h$ at point $m$:

$$w_h(m) = v_h(m)$$

since $m$ was arbitrary, $w_h \equiv v_h$.

**Lemma 3.** *Let $s_0 \in S$. The functions $w_s$, $s \neq s_0$ are linearly independent.*

PROOF. Assume a linear dependence between the $w_s$:

$$\sum_{s\neq s_0} \lambda_s w_s = 0$$

Then for any $a \in S$, $a \neq s_0$ (linearity of $\psi$):

$$0 = \psi\left(s_0, a; \sum_{s\neq s_0}\lambda_s w_s\right) = \gamma_a^{s_0 a}\gamma_a = \pm\lambda_a$$

hence $\lambda_a = 0$.

From the lemmas 1, 2, 3, we have:

**Theorem.** $\forall\, s_0 \in S$:

$$\{w_s, s\neq s_0\} \text{ is a basis of } J_h^1.$$

**Remark.**

(i) The above proof makes clear that a divergence free function $v_h \in V_h^1$ is entirely defined by the values of its fluxes through the paths between any couple of vertices.

(ii) Let $V_{h0}$ be the subset of functions in $V_h$, vanishing at boundary nodes. Then using the above technique the reader can check that $\{w_s,\ s \text{ interior vertex}\} \cup \{w_{\Gamma_i}\}$ is a basis of $V_{h0} \cap V_h^1$ as indicated in §3.2.1.

## Three Dimensional Case

F. Hecht (1980) has proved that an analogous construction can be performed in *3 dimensions* of space, in the case of a *simply connected* domain.[101] The elements are now *tetrahedra*; the discrete velocity is required to be:

Piecewise polynomial of degree $\leq 1$ per triangle;
Continuous at each face;
div $u_h = 0$ per element.

As above we decompose $V_h = V_h^1 \oplus V_h^2$:

$V_h^1$ (resp. $V_h^2$) $= \{v_h \in V_h | \text{for each face } F, v_h(M)\perp F \text{ (resp.} //F) \text{ at the}$ barycenter $M$ of the face$\}$

A basis of $V_h^2$ is easily found: for any face $F$, let $m$ be the barycenter of the face; choose any two independent vectors in the plane of the face: $t_{1m}, t_{2m}$.

---

[101] For example a toroidal domain is (at this moment) excluded.

Then two basis functions are associated to $m$, such that

at point $m$: $\mathbf{w}_{im}(m)=\mathbf{t}_{im}$, $m=1$ or $2$

at all other nodes: $\mathbf{w}_{im}(m')=0$

Then $\{\mathbf{w}_{im}\}$ is a basis of $V_h^2$.

The construction of a basis of $V_h^1$ is less straightforward. In the same way as we associated, in 2 dimensions a basis function to a vertex, we associate a function $\mathbf{w}_e$ to an edge as follows:

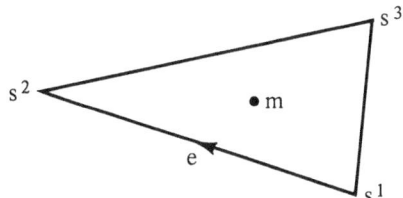

Choose an arbitrary orientation for the edge $e=(s^1, s^2)$;
If $(s^1 s^2 s^3)$ is one face with barycenter $m$, set:

$$\mathbf{w}_e(m)=2\frac{s^1 s^2 \wedge s^1 s^3}{|s^1 s^2 \wedge s^1 s^3|^2}$$

(so that $\mathbf{w}_e(m)$ is orthogonal to the face)

At all other mid face nodes (which do not share a face with the edge $e$), set:

$$\mathbf{w}_e(m)=0$$

$\mathbf{w}_e$ is a piecewise polynomial of degree $\leq 1$;
$\mathbf{w}_e$ is continuous at the mid face nodes.

We note that if: $e \subset$ face $F \subset T$, then

$$\left|\int_F \mathbf{w}_e|_T ds\right|=1$$

From the above remarks it is readily checked that:

$$\mathbf{w}_e \in V_h^1.$$

Now the trouble is we have too many functions to make a basis: this can be checked on simple examples with 1 or 2 tetrahedra. Thus we must remove some of the functions $\mathbf{w}_e$; let $E$ be the set of all the edges of the mesh, and $K$ the set of edges $e$ corresponding to the functions $\mathbf{w}_e$ to be *removed*. How should be $K$ so that $\{\mathbf{w}_e, e \in E \setminus K\}$ is a basis of $V_h^1$? The answer is remarkably simple, even though the proof is rather technical and uses a great deal of graph theory; $K$ may be constructed via the following algorithm ($N_E$=number of edges=$|E|$). The edges are numbered from 1 to $NE$.

**Algorithm 1:**

$$K: = \emptyset \text{ (void set)}$$
$$k: = 0$$
DO WHILE $k < NE$;
$$k: = k+1;$$
IF "$K \cup \{e_k\}$ does not contain any closed path"
THEN $K: = K \cup \{e_k\}$;
END;

(another algorithm will be given further below, that will be easier to implement).

In other words, $K$ is the greatest possible so that it does not contain any closed path.[102] Then, *if the domain is simply connected, the set $\{w_e, e \in E \setminus K\}$ is a basis of $V_h^1$.* The *proof* follows the same sketch as in 2D.

For any face $F$ with vertices $s^1 s^2 s^3$, let $\gamma$ be the closed path $s^1 s^2 s^3$ and choose an arbitrary orientation of the unit normal vector $\mathbf{n}$; define the flux of any function $\mathbf{v}_h \in V_h^1$ through as:

$$\psi(\gamma, \mathbf{v}_h) \equiv \int_F \mathbf{v}_h \cdot \mathbf{n} \, ds;$$

Extend this definition of the proof to any closed path; this part is the most technical of the proof.

Then

$$\psi(\gamma, \mathbf{w}_e) = \pm 1 \text{ if the edge } e \text{ is in the path } \gamma;$$
$$= 0 \text{ otherwise}$$

For any edge $e \in E \setminus K$, $K \cup \{e\}$ contains a closed path $\gamma_e$ (from the construction of $K$); for $\mathbf{v}_h \in J_h^1$, set

$$\mathbf{w}_h = \mathbf{v}_h - \sum_{e \notin K} \varepsilon_e \psi(\gamma_e, \mathbf{v}_h) \mathbf{w}_e$$

($\varepsilon_e = \pm 1$ depending upon the orientation of the edge $e$ in $\gamma_e$). Note that $\mathbf{w}_h \in V_h^1$.

It can be proved then, that: $\psi(\gamma^e, \mathbf{w}_h) = 0$ for any $e \in E \setminus K$, and from this: $\psi(\gamma, \mathbf{w}_h) = 0$, where $\gamma = s^1 s^2 s^3$ is the closed path associated to the face $s^1 s^2 s^3$. The last relation implies $\mathbf{w}_{h_1} \cdot \mathbf{n} = 0$ at the mid point of each face, that is, together with $\mathbf{w}_h \in V_h^1$, $\mathbf{w}_h \equiv 0$.

This proves that any $\mathbf{v}_h \in V_h^1$ can be generated by $\{\mathbf{w}_e, e \in E \setminus K\}$; the proof for the independency of this set is similar to the 2D case.

Now we have a basis of $V_h^1$, but another question arises, that of boundary conditions; in fact what we need is a basis of:

$$V_{h0}^1 = \{\mathbf{v}_h \in V_h^1 : \mathbf{v}_h = 0 \text{ at boundary nodes}\}$$

---

[102] Made of vertices joined by edges of the mesh; $K$ is a "tree" in terms of graph theory.

It can be proved (see again Hecht (1980)) that a basis of $V_{h0}^1$ is given by $\{w_e, \; e \in E \setminus H\}$, so that $H$ contains a maximum number of boundary edges.[103] The following algorithm constructs such a set $H$:

$EB$ = set of boundary edges
$EI$ = set of interior edges
$M$ = integer array, of dimension = number of vertices
$$= NV$$

**Algorithm 2**[104]

```
H: = ∅ (void set)
D   I = 1 TO NV;
M(I): = I
END;

FOR ALL e IN EB:DO;
CALL SP(e);
END;

FOR ALL e IN EI DO;
CALL SP(e);
END;

PROCEDURE SP(e);
I, J: = end-vertices of edge e;
IF M(I) ≠ M(J) THEN DO;
    H: + H ∪ {e} ; MJ: = M(J);
    LOOP : DO K = 1 TO, NE;
        IF M(K) = MJ THEN M(K): = M(I);
    END LOOP;
END SP;
```

**Remark.** At any step of the algorithm a graph is defined by $S$ (set of all vertices) and $H$ in its present state; then $M(I)$ assigns a number to the connected component of 1 in this graph;

$M(K) = M(I)$ for all $K$ in the same connected component as $I$.

For the treatment of nonhomogeneous boundary conditions we refer to Hecht (1980).

---

[103] The intuitive idea (following the 2D case) would remove *all* the boundary edges, but this appears to be a pitfall.

[104] These algorithms are displayed in a "pseudo-ALGOL" notations which we hope to be self-explanatory.

# References

## Abbreviations:

*Comp. Meth. Appl. Mech. Eng.: Computer Methods in Applied Mechanics and Engineering,* Argyris, T. H., Prager, W., and Holt, M., eds. North-Holland Pub. Co., Amsterdam.
*Int. J. Num. Meth. Engng.: International Journal for Numerical Methods in Engineering,* Zienckiewicz and Gallagher, eds.
*J. Comp. Ph: Journal of Computational Physics.*
*R.A.I.R.O.: Revue Française d'Automatique, d'Informatique et de Recherche Operationnelle,* Dunod, Paris.
*C.R.A.S.: Comptes Rendus des Séances de l'Académie des Sciences,* Paris.
*FENOMECH'78: Proceedings of Int. Conf. on Finite Elements in Nonlinear Mechanics,* held in Universitat Stuttgart, pub. in Comp. Meth. Appl. Mech. Eng., 17/18 (March 1979).

Absi, E. and Glowinski R., eds. (1979), "Methodes Numeriques dans les Sciences de l'Ingenieur," *Proceedings of: 1er congres du G.A.M.N.I.* (1978), Dunod, Paris.
Agmon, S. (1965), *Lectures on Elliptic Boundary Value Problems,* Van Nostrand, New York.
Atkins, D. J., Maskell, S. J., and Patrick M. A. (1980), "Numerical prediction of separated flows," *Int. J. Num. Meth. Engng.* 15 (1980), pp. 129–144.
Atkinson, J. D. and Hughes T. J. R. (1977), "Upwind Finite Element schemes for Convective-Diffusive Equations," technical note C-2, Dec. 1977, Charles Kolling Res. Lab., The University of Sydney, N.S.W., Australia.
Aufaure, M. and Benjamin, C. (1970), "Application de la méthode des éléments finis a la résolution numérique des problèmes d'élasticité," *Revue Française de Mécanique,* 33 (1970), pp. 5–18.
Aussems, A. (1972), "Une mise en oeuvre frontale des méthodes d'éléments finis," thèse, Université Scientifique et Médicale de Grenoble, France.
Axelsson, O. (1976), "A class of iterative methods for finite element equations," *Comp. Meth. Appl. Mech. Eng.,* 9 (1976), 2, pp. 123, 138.
Axelsson, O. and Gustafsson, I. (1979), "An iterative solver for a mixed variable variational formulation of the (first) biharmonic problem," *Comp. Meth. Appl. Mech. Eng.,* 20 (1979), pp. 9–16.
Aziz, A. K., ed. (1972), *The Mathematical foundations of the Finite Element Method, with applications to Partial Differential Equations,* Academic Press, New York and London.
Babuska, I. (1973), "The finite element method with penalty," *Math. of Comp.,* 27 (April 1973), no. 122, pp. 221–228.
Babuska, I. and Aziz, A. K. (1972), "Survey Lectures on the Mathematical foundations of the Finite Element Method," pp. 3–359, in Aziz (1972).
Babuska, I. and Aziz, A. K. (1976), "On the angle condition in the finite element method," *SIAM J. Numer. Anal.,* 13 (1976), no. 2, pp. 214–226.
Babuska, I. and Rheinboldt, W. C. (1978), "Adaptive approaches and reliability estimations in Finite Element Analysis," *in Proceedings of FENOMECH'78.*
Baker, A. J. (1973), "Finite element solution algorithms for viscous incompressible fluid dynamics," *Int. J. Num. Meth. Eng.,* 6 (1973), no. 1, pp. 89–101.
Baker, A. J. and Soliman, M. O. (1979), "Utility of a finite element solution algorithm for initial value problems," *J. Comp. Ph.,* 32 (1979), pp. 289–324.

Bardos, C., Bercovier, M., and Pironneau, O. (1980), "The Vortex method with finite elements," Rapport de recherche INRIA no. 15, Mars 1980.

Barrett, K. E. (1974), "The numerical solution of singular perturbation Boundary value problems," *Quart. J. Mech. and Appl. Math.*, **27** (1974), pp. 57–68.

Barrett, K. E., and Demunshi, G. (1979), "Finite element solutions of convective diffusion problems," *Int. J. Num. Meth. Eng.*, **14** (1979), pp. 1511–1524.

Becker, E. B., and Carey, G. F. (1978), "Adaptive Mesh Refinement and Non Linear fluid mechanics," in *Proceedings of FENOMECH'78*.

Begis, D. (1979), "Etude numerique de l'écoulement d'un fluide viscopiastique de Bingham par une methode de Lagrangien augmenté," Rapport de Recherche No. 355, LABORIA-IRIA.

Begis, D. and Perronnet, A. (1980, to appear), *The Club MODULEF*, Dunod, Paris.

Benazeth, J. C. (1978), "Résolution des équations de Navier-Stokes stationnaires par une méthode d'éléments finis mixtes," thèse de 3eme. cycle, Université de Paris 6.

Benjamin, A. S. and Denny, V. E. (1979), "On the convergence of numerical solutions for 2-D flows in a cavity at large Reynolds number," *J. Comp. Ph.*, **33** (1979), pp. 340–358.

Benque, J. P., Ibler, B., Keraimsi, A., and Labadie, G. (1979), "A Finite Element Method for Navier-Stokes equations," Rapport EDF-DER-LNH, in Norrie (1980), vol. 1, pp. 110–120.

Bercovier, M. (1976), "Regularisation duale des problemes variationnels mixtes. Application aux éléments finis mixtes, et extension à quelques problemes non lineaires," These de Doctorat d'Etat, Universite de Rouen, France.

Bercovier, M. (1977), "A family of finite elements with penalisation for the numerical solution of Stokes and Navier-Stokes equations," in Gilchrist (1977).

Bercovier, M. (1978), "Perturbation of Mixed variational problems, Applications to mixed finite element methods," *R.A.I.R.O.*, **12** (1978), pp. 211–236.

Bercovier, M. and Engelman, M. (1979), "A finite element for the numerical solution of viscous incompressible flows," *J. Comp. Physics*, **30** (1979), pp. 181–201.

Bercovier, M. and Pironneau, O. (1977), "Estimations d'erreur pour la résolution du problème de Stokes en éléments finis de Lagrange," *C.R.A.S.*, Série A-555, 14 novembre 1977.

Bercovier, M. and Pironneau, O. (1979), "Error estimates for Finite Element Method solution of the Stokes problem in the primitive variables," *Numerische Mathematik*, **33** (1979), pp. 211–224.

Bernardi, C. (1979), "Methode d'éléments finis mixtes pour les équations de Navier Stokes," Thèse de doctorat d'état, Université Paris 6.

Boisvert, E. (1979), "Méthode de triangulation automatique et de pénalisation pour les équations de Stokes et Navier-Stokes," Thèse de 3ème cycle, Université Paul Sabatier, Toulouse, France.

Bolley, C. and Crouzeix, M. (1978), "Conservation de la positivité lors de la discrétisation des problèmes d'évolution paraboliques," *R.A.I.R.O.*, **12** (1978), no. 3, pp. 237–245.

Bonnet, M. and Meurant, G. (1979), "Comparaison de différentes méthodes itératives de résolution des systèmes linéaires et non linéaires," in Absi and Glowinski (1979).

Bonnet, M. and Perronnet, A. (1973), "Décompositions du problème de Stokes, stationnaire et évolutif, par dualisation de certaines contraintes," Thèse de 3ème cycle, Université Paris 6, Laboratoire d'analyse numérique.

Boor, C. de, ed. (1974), *Mathematical Aspects of Finite Element Methods in Partial Differential Equations*, Academic Press, New York.

Borrel, M. (1977a), "Résolution des équations des équations de Navier-Stokes par une méthode d'éléments finis," Technical note Onera no. 1977/8.

Borrel, M. (1977b), "Etude et mise en oeuvre d'une méthode d'éléments finis de résolution des équations de Navier-Stokes," Thèse de 3ème cycle, Université Paris 6.

Bossavit, A. (1971), "Une méthode de décomposition de l'opérateur biharmonique," note HI/585/2, Electricité de France.

Bossavit, A. and Fremond, M. (1976), "The frontal method based on mechanics and dynamic programming; an algebraic account," *Comp. Meth. App. Mech. Eng.*, **8** (1976), pp. 153–178.

Bourgat, J. F. (1971), "Analyse numérique du problème de la torsion élastoplastique," Thèse de 3ème cycle, Université Paris 6.

Bourgat, J. F. (1976), "Numerical applications of a dual iterative method for solving a finite element approximation of the biharmonic equation," *Comp. Meth. Appl. Mech. Eng.*, **9**, (1976), pp. 203–218, and Rapport de recherche no. 156 (LABORIA-IRIA).

Bozeman, J. D. and Dalton, C. (1973), "Numerical study of viscous flow in a cavity," *J. Comp. Ph.*, **12** (1973), pp. 348–363.

Bramellan, A., Allen, R. N., and Harman, Y. N. (1976), *Sparsity*, Pitman Publishing, London.

Bredif, M. (1980), "Résolution des équations de Navier-Stokes par éléments finis mixtes," Thèse de docteur-ingénieur, U. Paris 6, June 1980.

Brezzi, F. (1974), "On the existence, uniqueness and approximation of saddle point problems arising from Lagrange multipliers," *R.A.I.R.O.*, série rouge, **R2**, pp. 129–151.

Brezzi, F. and Marini, L. D. (1975), "On the numerical solution of plate bending problems by hybrids methods," *R.A.I.R.O.*, **R-3** (1975), pp. 5–50

Brezzi, F. and Raviart, P. A. (1978), "Mixed finite element methods for fourth order elliptic equations," in Miller (1978).

Brezzi, F., Chinosi, C., Della Croce, L., Marini, L. D., Quarteroni, A., Sacchi, G., and Scapolla, T. (1979), "Recent developments in non-standard Finite Element Methods," pp. 3–11, in Absi and Glowinski (1979).

Bristeau, M. O. (1975), "Application de la méthode des éléments finis à la résolution numérique d'inéquations variationnelles de type Bingham," Thèse de 3ème cycle, Université Paris 6 (Juin 1975).

Bristeau, M. O. (1977), "Application of optimal control theory to transonic flow computations by Finite Element methods," in Glowinski and Lions (1977).

Bristeau, M. O. (1978), "Application of a finite element method to transonic flow problems using an optimal control approach," VKI Lecture series, (Computational Fluid Dynamics, March 13–17, 1978, Rhode Ste. Genèse, Belgium).

Bristeau, M. O., Glowinski, R., and Pironneau, O. (1977), "Numerical solution of the transonic equation by the finite element method via optimal control," in *Control Theory of Systems Governed by Partial Differential Equations*, Academic Press, New York, (1977).

Bristeau, M. O., Glowinski, R., Periaux, J., Perrier, P., and Pironneau, O. (1979), "On the numerical solution of non linear problems in fluid dynamics by least squares and finite element methods," in *Proceedings of FENOMECH'78, Comp. Meth. Appl. Mech. Eng.*, **17/18**, (1979), part 3, pp. 619–657.

Bristeau, M. O., Glowinski, R., Periaux, J., Perrier, P., Pironneau, O. and Poirier, G. (1978), "Application of optimal control and finite element methods to the calculation of transonic flows and incompressible viscous flows," Rapport de recherche no. 294 (avril 1978, LABORIA-IRIA).

Bristeau, M. O., Glowinski, R., Mantel, B., Periaux, J., Perrier, P., and Pironneau, O., (1979), "A Finite Element approximation of Navier-Stokes equations for incompressible viscous fluids," in Rautmann (1980).

Brodlie, K. W., Gourlay, A. R., and Greenstadt, J. (1973), "Rank-one and rank-two corrections to positive definite matrices expressed in product form," *J. Inst. Math. Appl.*, **11** (1973), pp. 73–82.

Bunch, J. R. and Rose, D. J., eds. (1976), "Sparse matrix computations," *Proceedings of Symposium on Sparse Matrix Computations*, held at Argonne Nat. Lab., in Sept. 1975, Academic Press.

Burggraf, O. R. (1966), "Analytical and numerical studies of the structure of steady separated flows," *J. Fluid Mech.*, **24** (1966), pp. 113–151.

Campion-Renson, A. and Crochet, M. J. (1978), "On the stream function vorticity finite element solutions of Navier-Stokes equations," *Int. J. Num. Meth. Eng.*, **12** (1978), pp. 1809–1818.

Cantin, G. (1971), "An equation solver of very large capacity," *Int. J. Num. Meth. Eng.*, **3** (1971), pp. 379–388.

Cavendish, J. C. (1974), "Automatic triangulation of arbitrary planar domains for the finite element method," *Int. J. Num. Meth. Eng.*, **8** (1974).

Chattot, J., Guiu, J., and Laminie, J. in Norrie (1980)

Chauvet, Y. (1979), "Resolution de l'équation instationnaire du transport par une méthode d'éléments finis discontinus," in Absi and Glowinski (1979).

Chavent, G. (1979), "Finite element method for the water flooding problem," in Glowinski and Lions (1979).

Cheung, Y. K. and Khatua, T. P. (1976), "A finite element solution program for large structures," *Int. J. Num. Meth. Eng.*, **10** (1976), pp. 401–412.

Chorin, A. J. (1967), "A numerical method for solving incompressible viscous flow problems," *J. Comp. Ph.*, **2** (1967), pp. 12–26.

Chorin, A. J. (1968a), "Numerical solution of incompressible flow problems," *Studies in Numer. Anal.*, **2** (1968), pp. 64–71.

Chorin, A. J. (1968b), "On the convergence and approximation of discrete approximations to the Navier-Stokes equations," *Math. of Comp.*, **23** (1968), pp. 341–353.

Chorin, A. J. (1973), "Numerical study of slightly viscous flow," *J. Fluid Mech.*, **57** (1973), pp. 785–796.

Christie, I., Griffiths, D. F., Mitchell, A. R., and Zienckiewicz, O. C. (1976), "Finite element methods for second order differential equations with significant first derivatives," *Int. J. Num. Meth. Eng.*, **10** (1976), pp. 1389–1396.

Chung, T. J. (1978), *Finite Element Analysis in Fluid Dynamics*, McGraw-Hill, New York.

Ciarlet, P. G. (1975), "Numerical analysis of the finite element method," *Séminaire de Mathématiques Supérieures*, Presses de l'Université de Montréal.

Ciarlet, P. G. (1978), *The Finite Element Method for Elliptic Problems*, North-Holland, Amsterdam.

Ciarlet, P. G. and Glowinski, R. (1975), "Dual iterative techniques for solving a finite element approximation of the biharmonic equation," *Comp. Meth. Appl. Mech. Eng.*, **5** (1975), pp. 277–295.

Ciarlet, P. G. and Raviart, P. A. (1972a), "Interpolation theory over curved elements, with applications to finite elements methods," *Comp. Meth. Appl. Mech. Eng.*, **1** (1972), pp. 217–249.

Ciarlet, P. G. and Raviart, P. A. (1972b), "General Lagrange and Hermite interpolation with applications to finite element methods," *Arch. Rat. Mech. Anal.*, **46** (1972), pp. 177–199.

Ciarlet, P. G. and Raviart, P. A. (1972c), "The combined effect of curved boundaries and numerical integration in isoparametric finite element methods," pp. 409–474, in Aziz (1972).

Ciarlet, P. G. and Raviart, P. A. (1973), "Maximum principle and uniform convergence for the finite element method," *Comp. Meth. Appl. Mech. Eng.*, **2** (1973), pp. 17–31.

Ciarlet, P. G. and Raviart, P. A. (1974), "A mixed finite element method for the biharmonic equation," in de Boor (1974).

Concus, P. and Golub, G. H. (1973), "Use of fast direct methods for the efficient numerical solution of non separable elliptic equations," *SIAM J. Numer. Anal.*, **10** (1973), pp. 1103–1120.

Concus, P. and Golub, G. H. (1975), "A generalized conjugate gradient method for nonsymmetric systems of linear equations," in Glowinski and Lions (1975).

Cooke, C. H. (1978), "A split band Cholesky equation solving strategy for finite element analysis of transient field problems," *Int. J. Numer. Meth. Eng.*, **12** (1978), pp. 703–710.

Courant, R. (1943), "Variational methods for the solution of problems of equilibrium and vibrations," *Bull. Amer. Math. Soc.*, **49** (1943), pp. 1–23.

Crisfield, M. A. (1979), "A faster modified Newton Raphson Iteration," *Comp. Meth. Appl. Mech. Eng.*, **20** (1979), pp. 267–278.

Critescu, C. and Loubignac, G. (1979), "Quadratures de Gauss pour des fonctions avec singularité en $1/r$ sur des carrés ou des rectangles," pp. 21–32, in Absi and Glowinski (1979).

Crouzeix, M. (1974), "Etude d'une méthode de linéarisation. Résolution numérique des équations de Stokes stationnaires. Application aux équations de Navier-Stokes stationnaires," *Cahiers de l'IRIA*, no. 12.

Crouzeix, M. (1976), *Proceedings of: Journées éléments finis*, Université de Rennes.

Crouzeix, M. and Raviart, P. A. (1973), "Conforming and non-conforming finite element methods for solving the stationary Stokes equations," *R.A.I.R.O.*, **R-3** (1973), pp. 33–76.

Crouzeix, M. and Thomas, J. M. (1973), "Elements finis et problèmes elliptiques dégénérés," *R.A.I.R.O.*, **R-3**, Déc. 1973, pp. 77–104.

Curtis, A. R. and Reid, J. K. (1971), "The solution of large sparse unsymmetric systems of linear equations," *J. Inst. Math. Appl.*, **8** (1971), pp. 344–353.

Cuthill, E. and Mc. Kee, J. M. (1969), "Reducing the bandwidth of sparse symmetric matrices," *Proceedings of 24th. Nat. Conf., Assoc. Comp. Mach.*, ACM pub. p. 69.

Daniel, J. W. (1971), *The Approximate Minimization of Functionals*, Academic Press, New York.

Datta, A. B. and Strauss, K. (1976), "Slow flow of a viscoelastic fluid through a contraction," *Rheol. Acta*, **15** (1976), pp. 403–410.

Dennis, S. C. R., Ingham, D. B., and Cook, R. N. (1979), "Finite difference methods for calculating steady incompressible flows in three dimensions," *J. Comp. Ph.*, **33** (1979), pp. 325–339.

Dervieux, A. and Thomasset, F. (1979), "A finite element method for the simulation of a Rayleigh-Taylor instability," in Rautmann (1980).

Douglas, Jr., J. (1972), "A super-convergence result for the approximate solution of the heat equation by a collocation method," in Aziz (1972).

Douglas, Jr., J. and Dupont, T. (1972), "Some superconvergence results for Galerkin methods for the approximate solution of the two-point boundary value problem," *Proceedings of the Conference on Numerical Analysis*, Royal Irish Academy, Dublin, Published by J. H. Miller.

Duff, I. S. (1979), "Some current approaches to the solution of large sparse systems of linear equations," in Absi and Glowinski (1979).

Duvaut, D., and Lions, J. L. (1971), *Les Inéquations en Mécanique et en Physique*, Dunod, Paris.

Ehrlich, L. W. (1971), "Solving the biharmonic equations as coupled difference equations," *SIAM J. Num. Anal.*, **8** (1971), no. 2, pp. 278–287.

Engelman, M. S. (1979), "Computer simulation of flow models of newtonian and non-newtonian viscous incompressible flows," Thesis, The Hebrew University of Jerusalem.

Engquist, B. and Kreiss, H. O. (1979), "Difference and finite element methods for hyperbolic differential equations," *Comp. Meth. Appl. Mech. Eng.*, **17/18** (1979), pp. 581–596.

Felippa, Carlos A. (1978), "Iterative procedures for improving penalty function solutions of algebraic systems," *Int. J. Num. Meth. Eng.*, **12** (1978), pp. 821–836.

Figueroa, J. (1979), "La methode d'éléments finis de Hermann Johnson pour les équations de Navier-Stokes," thèse de 3ème cycle, Université Paris 6, and Centre de Mathématiques appliquées, Ecole Polytechnique, Palaiseau.

Fix, G. (1972), "Effects of quadrature errors in finite element approximation of steady state, eigenvalue and parabolic problems," pp. 525–556, Aziz (1972).

Fletcher, C. A. J. (1979), "A primitive variable finite element formulation for inviscid, compressible flow," *J. Comp. Ph.*, **33** (1979), pp. 301–312.

Forsythe, G. E. and Wasow, W. (1960), *Finite Difference Methods for Partial Differential Equations*, Wiley, New York.

Fortin, M. (1972), "Calcul numérique des écoulements des fluides de Bingham et des fluides newtoniens incompressibles par la méthode des éléments finis," Thèse, Université Paris 6.

Fortin, M. (1976), "Résolution numérique des équations de Navier-Stokes par des éléments finis de type mixte," Rapport de Recherche 184 (LABORIA-IRIA), Aout 1976.

Fortin, M. (1977), "Analysis of the convergence of mixed finite element methods," *R.A.I.R.O., Série Analyse Numérique*, **11** (1977), pp. 344–354.

Fortin, M. and Glowinski, R., eds. (1980, to appear), *Numerical Solution of Boundary Value Problems by Augmented Lagrangians*, Dunod, Paris.

Fortin, M. and Thomasset, F. (1979), "Mixed finite element methods for incompressible flow problems," *J. Comp. Ph.*, **31** (1979), pp. 113–145.

Fraeijs de Veubeke, B. (1965), "Bending and stretching of plates," in *Conf. on Matrix Meth. In Structural Mech.*, Wright Patterson, AFB, Ohio.

Fritts, M. J. and Boris, J. P. (1979), "The lagrangian solution of transient problems in hydronamics using a triangular mesh," *J. Comp. Ph.*, **31** (1979), pp. 173–215.

Gallagher, R. H. (1975), *Finite Element Analysis: Fundamentals*, Prentice-Hall, Englewood Cliffs, New Jersey.

Gallagher, R. H., Oden, J. T., Taylor, C., and Zienckiewicz, O. C., eds. (1975), *Proceedings of the first International Conference on Finite Element Methods in Flow Problems* (Swansea, 1974), The University of Alabama Press, 1975.

Gallagher, R. H., Morandi-Cecchi, M., Oden, J. T., Taylor, C., and Zienckiewicz, O. C., eds. (1977), *Proceedings of the Second Int. Conf. on F.E.M. in Flow problems*, (Santa Margherita Ligure, 1976).

Gallagher, R. H., Oden, J. T., Taylor, C., and Zienckiewicz, O. C., eds. (1975–1979), *Finite Elements in Fluids*, Vol. 1 and 2 (1975); Vol. 3 (1979), Wiley, New York.

Gallagher, R. H., Yamada, Y., and Oden, J. T., eds. (1971), *Recent Advances in Matrix Methods of Structural Analysis and Design*, The University of Alabama Press.

Galligani, I. and Magenes, E., eds. (1977), "Mathematical aspects of finite element methods," *Lecture Notes in Mathematics* 606, Springer-Verlag, New York.

Gartling, D. K. and Becker, E. B. (1976), "Finite element analysis of viscous incompressible flow," *Comp. Meth. Appl. Mech. Eng.*, **8** (1976), pp. 51–60 and 127–138.

Gartling, D. K. (1978), "Some comments on the paper by Heinrich, Huyakorn, Zienckiewicz and Mitchell," *Int. J. Num. Meth. Eng.*, **12** (1978), pp. 187–190.

George, J. A. (1971), "Computer implementations of the finite method," thesis, Comp. Sc. Dept., report no. STAN/CS/71/208, Stanford University.

George, J. A. (1973), "Nested dissection of a regular finite element mesh," *SIAM J. Numer. Anal.*, **10** (1973), pp. 345–363.

George, J. A. (1977), "Sparse matrix aspects of the finite element method," in Glowinski and Lions (1977).

Gilchrist, B. ed. (1977), *Proceedings of IFIP Symposium*, North-Holland, Amsterdam.

Girault, V. (1976), "A combined finite element and MAC method for solving the Navier-Stokes equations," *Numerische Mathematik*, **26** (1976), pp. 39–59.

Girault, V. and Raviart, P. A. (1979), "Finite element approximation of the Navier-Stokes equations," *Lecture Notes in Mathematics 749*, Springer-Verlag, New York.

Glowinski, R. (1973), "Approximations externes par éléments finis d'ordre 1 et 2 du problème de Dirichlet pour l'opérateur biharmonique. Méthode itérative de résolution des problèmes approachés," in *Topics in Numerical Analysis*, J. Miller ed., Academic Press, pp. 123–171.

Glowinski, R. and Lions, J. L., eds. (1973, 1975, 1977, 1979), "Computing methods in applied sciences and engineering," *Proceedings of international symposiums organized by IRIA-LABORIA*: 1st. symposium, déc. 1973, pub. by Springer-Verlag, Lecture notes in Computer Science (1974): part 1, no. 10; part 2, no. 11; 2nd. symposium, déc. 1975, Springer-Verlag, Lecture notes Comp. Sc., (1976) 58; 3rd. symposium, déc. 1977, Springer-Verlag, Lecture notes Comp. Sc., (1978) 134; 4th. symposium, déc. 1979, pub. by North-Holland, Amsterdam, (1980, to appear).

Glowinski, R., Mantel, B., Periaux, J., Pironneau, O., and Poirier, G. (1979), "An efficient preconditioned conjugate gradient method applied to non linear problems in fluid dynamics," in Glowinski and Lions (1979).

Glowinski, R., Mantel, B., Periaux, J., Perrier, P., and Pironneau, O. (1980, to appear), "Applications of non linear least squares methods to non linear problems in fluiddynamics," to appear in *Proceedings of Int. Conf. on Finite el. Meth.*, Calgary.

Glowinski, R., Periaux, J., and Pironneau, O. (1976), "Transonic flow simulation by the finite element method, via optimal control," in Gallagher, Morandi-Cecchi *et. al.* (1976).

Glowinski, R., Periaux, J., and Pironneau, O. (1978–1979), "An efficient preconditioning scheme for iterative numerical solutions of partial differential equations," pré-publications mathématiques, Université Paris-Nord, Dept. de Maths.

Glowinski, R. and Pironneau, O. (1976), "On the computation of transonic flows," in *Proceedings of the 1st Franco-Japanese Colloquium on Functional and Numerical Analysis*, (Tokyo, Sept. 1976).

Glowinski, R. and Pironneau, O. (1977), "Numerical methods for the first biharmonic equation and for the two-dimensional Stokes problem," Comp. Sc. Dept., report STAN/CS/77/615, Stanford University, May 1977.

Glowinski, R. and Pironneau, O. (1978a), "Approximation par éléments finis mixtes du problème de Stokes en formulation vitesse-pression; convergence des solutions approchées," *C.R.A.S.* Paris, T.268A (1978), pp. 181–183.

Glowinski, R. and Pironneau, O. (1978b), "Approximation par éléments finis mixtes du problème de Stokes en formulation vitesse-pression; résolution des problèmes approchées," *C.R.A.S.* Paris, T268A (1978), pp. 225–228.

Glowinski, R. and Pironneau, O. (1979a), "On numerical methods for the Stokes problem," ch. 13, pp. 243–264, in Glowinski *et. al.* (1979).

Glowinski, R. and Pironneau, O. (1979b), "On a mixed finite element approximation of the Stokes problem," *Numerische Mathematik*, **33** (1979), pp. 397–424.

Glowinski, R. and Pironneau, O. (1979c), "Numerical methods for the biharmonic equation and for the two dimensional Stokes problem," *SIAM Review*, **17** (1979), no. 2, pp. 167–212.

Glowinski, R., Rodin, E. Y., and Zienckiewicz, O. C., eds. (1979), "Energy methods in finite element analysis," Wiley, New York.

Gresho, P. M., Lee, R. L., and Sani, R. L. (1976), "Advection dominated flows with emphasis on the consequences of mass-lumping," in Gallagher *et. al.* (1976).

Griffiths, D. F. (1976), "On the approximation of convection problems in fluid dynamics," Res. paper no. 317, Sept. 1976, Dept. of Math. and Stat., The University of Calgary, Alberta, Canada.

Grooms, H. R. (1972), "Algorithms for matrix bandwidth reduction," *J. Struct. Div., ASCE*, **98**, Proc. paper 8636, pp. 203–214.

Guillaume, P. (1980), Thèse de 3ème. cycle, Université Paris 11 (Orsay).

Gupta, M. M. and Manohar, Ram P. (1979) "Boundary approximations and accuracy in viscous flow computations," *J. Comp. Ph.*, **31** (1979), pp. 265–288.

Hasbani, Y. and Engelman, M. (1979), "Out of core solution of linear equations with non symmetric coefficient matrix," *Computers and Fluids*, **7** (1979), pp. 13–31.

Hecht, F. (1980), "Construction d'une base de fonctions p1 non conformes à divergence nulle dans R3," *R.A.I.R.O.*, to appear (1980).

Heinrich, J. C., Huyakorn, P. S., Zienckiewicz, O. C., and Mitchell, A. R. (1977), "An upwind finite element scheme for the two-dimensional convective equation," *Int. J. Num. Meth. Eng.*, **11** (1977), pp. 131–143.

Heinrich, J. C. and Zienckiewicz, O. C. (1977), "Quadratic finite element scheme for two dimensional convective transport problems," *Int. J. Num. Meth. Eng.*, **11** (1977), pp. 1831–1844.

Hestenes, M. R. (1969), "Multiplier and gradient methods," J.O.T.A., **4** (1969), pp. 303–320.

Hestenes, M. R. and Stiefel, E. (1952), "Methods of conjugate gradients for solving linear systems," *J. Res. Nat. Bur. Stand.*, **49** (1952), pp. 409–436.

Himmelblau, D. M., ed. (1973), "Decomposition of large scale problems," North-Holland, Amsterdam.

Hinton, E. and Owen, D. J. R. (1977), *Finite Element Programming*, Academic Press, New York.

Hood, P. (1976), "Frontal solution program for unsymmetric matrices," *Int. J. Num. Meth. Eng.* **10** (1976), pp. 379–399.

Hua, Bach Lien and Thomasset, F. (1979a), "Modélisation numérique de phénomènes d'"upwelling" cotiers par une méthode d'éléments finis non conformes," rapport de recherche 366 (LABORIA-IRIA), Oct. 1979.

Hua, Bach Lien and Thomasset, F. (1979b), "Numerical study of coastal upwellings by a finite element method," in Glowinski and Lions (1979).

Hughes, T. J. R. (1978), "A simple finite element scheme for developping upwind finite elements," *Int. J. Num. Meth. Eng.*, **12** (1978), pp. 1359–1365.

Hughes, T. J. R., Taylor, R. L., and Levy, J. F. (1976), "A finite element scheme for incompressible viscous flows," in Gallagher, Morandi-Cecchi *et. al.* (1976); also in Gallagher *et. al.* (1979, vol. 3).

Hughes, T. J. R., Wing Kam Liu, and Brooks, A. (1979), "Finite element analysis of incompressible viscous flows by the penalty function formulation," *J. Comp. Ph.*, **30** (1979), pp. 1–40.

Hughes, T. J. R., Wing Kam Liu, and Zimmermann, T. K. (1978), "Lagrangian eulerian finite element formulation for incompressible viscous flows," proceedings of U.S.-Japan Conf. on Interdisciplinary finite element analysis, Cornell University, August 7–11, 1978.

Hutton, S. G., Exeter, M. K., Fussey, D. E., and Webster, J. J. (1980), "Primitive variable finite element formulations for steady viscous flows," *Int. J. Num. Meth. Eng.*, **15** (1980), pp. 209–223.

Huyakorn, P. S. (1977), "Solution of the steady state convective transport equation using an upwind finite element scheme," *Appl. Math. Modelling*, **1** (1977), p. 187.

Huyakorn, P. S., Taylor, C., Lee, R. L., and Gresho, P. M. (1978), "A comparison of various mixed interpolation finite elements for the Navier-Stokes equations," *Computer and Fluids*, **6** (1978), pp. 25–35.

Ikegawa, M. (1979), "A new finite element technique for the analysis of viscous flow problems," *Int. J. Num. Meth. Eng.*, **14** (1979), pp. 103–113.

Ikenouchi, M. and Kimura, N. (1974), "An approximate numerical solution of the Navier-Stokes equations by Galerkin methods," in Gallagher, Oden, Taylor, and Zienckewicz (1975), pp. 99–100.

Irons, B. M. (1970), "A frontal solution program for finite element analysis," *Int. J. Num. Meth. Eng.*, **2** (1970), pp. 5–32.

Irons, B. M. and Razzaque, A. (1972), "Experiences with the patch test for convergence of finite elements," in Aziz (1972).

Jaffre, J. (1979a), "Approximation par une méthode d'éléments finis mixtes d'une équation du type diffusion convection stationnaire," Rapport de recherche 367 (LABORIA-IRIA), Oct. 1979.

Jaffre, J. (1979b), "Approximation of a diffusion convection equation by a mixed finite element method; application to the water flooding problem," *Proceedings of TICOM Conf.*, Austin, Texas, to appear in Computers and Fluids.

Jamet, P. (1976), "Estimation d'erreur pour des éléments finis droits presque dégénérés,"
   *R.A.I.R.O.*, série Analyse numérique, **10**, no. 3, mars 1976, pp. 43–61.
Jamet, P. (1975), "Estimation d'erreur pour des éléments finis quadrilatéraux de type Q1, qui
   peuvent dégénérer en triangles," note *C.R.A.S.*, tome 281, déc. 1975, série A, pp. 983–984.
Jamet, P. and Raviart, P. A. (1973), "Numerical solution of the stationary Navier-Stokes
   equations by finite element methods," tome 1, pp. 193–223, in Glowinski and Lions (1973).
Jennings, A. (1966), "A compact storage scheme for the solution of symmetric linear simulta-
   neous equations," the *Computer J.*, **9** (1966), pp. 281–285.
Jennings, A. and Tuff, A. D. (1971), "A direct method for the solution of large sparse symmetric
   simultaneous equations," in *Large Sparse Sets of Linear Equations*, J. K. Reid, ed.,
   Academic Press.
Jennings, A. and Tuff, A. D. (1973), "An iterative method for large systems of linear structural
   equations," *Int. J. Num. Meth. Eng.*, **7** (1973), pp. 175–183.
Johnson, C. (1978), "A mixed finite element method for the Navier-Stokes equation," *R.A.I.R.O.*,
   *Série Analyse Numérique*, **12** (1978), no. 4, pp. 335–348.
Johnson, C. and Mercier, B. (1978), "Some equilibrium finite element methods for two dimen-
   sional elasticity problems," *Numerische Mathematik*, **30** (1978), no. 1.
Joly, P. (1977), "La méthode frontale," Rapport Interne 77/008, Laboratoire d'Analyse
   numérique (LA189), Université Paris 6.
Jones, R. (1973), "QMESH: A self-organizing mesh generation program," technical report, no.
   SLA/73/1088, Sandia Lab., Albuquerque, New Mexico.
Kershaw, D. S. (1978), "The Incomplete Cholesky-conjugate gradient method for the iterative
   solution of systems of linear equations," *J. Comp. Ph.*, **26** (1978), pp. 43–65.
Kikuchi, F. and Ushijima, T. (1980), "Theoretical analysis of some finite element schemes for
   convective diffusion equations," in Norrie (1980), vol. 1, pp. 82–95.
Kopal, Z. (1961), *Numerical Analysis*, Chapman and Hall, London.
Koutchmy, O., Joly, P., and Perronnet, A. (1977), "Les modules de maillage bidimensionnels du
   club MODULEF," Université Paris 6, Laboratoire d'Analyse Numérique (LA189).
Kreiss, H. O. (1979), "Numerical methods for hyperbolic partial differential equations," in
   Computational Fluid Dynamics, *VKI Lecture Series 1979/6*.
Lailly, P. (1976), "Résolution numérique des équations de Stokes en symétrie de révolution par
   une méthode d'éléments finis non conformes," Thèse de Docteur-Ingénieur, Université Paris
   11 (Orsay).
Lascaux, P. (1976), "Numerical methods for time dependent equations. Applications to fluid
   flow problems," Tata Inst. of Fundamental Res., Bombay, India.
Lascaux, P. and Lesaint P. (1975), "Some non conforming finite elements for the plate bending
   problem," *R.A.I.R.O.*, série analyse numérique, **9** (1975), pp. 9–53.
Lee, R. L., Gresho, P. M., and Sani, R. L. (1979), "Smoothing techniques for certain primitive
   variable solutions of the Navier Stokes equations," *Int. J. Num. Meth. Eng.*, **14** (1979), pp.
   1785–1804.
Legait, B. (1980), "Contribution a l'étude numérique des équations de Navier-Stokes à deux
   phases," Institut Français du Pétrole, rapport no. 28247, Juilliet 1980.
Leroux, A. Y. (1979), "Approximation de quelques problèmes hyperboliques non linéaires,"
   Thèse de Doctorat d'Etat, Université de Rennes.
Lesaint, P. (1975), "Sur la résolution des systèmes hyperboliques du premier ordre par des
   méthodes d'éléments finis," Thèse de doctorat d'état, Université Paris 6.
Lesaint, P. and Raviart, P. A. (1974), "On a finite element method for solving the neutron
   transport equation," pp. 89–123, in C. de Boor (1974).
Letallec, P. (1978), "Simulation numérique d'écoulements visqueux incompressibles par des
   méthodes d'éléments finis mixtes," Thèse de 3ème. cycle, Université Paris 6, Laboratoire
   d'Analyse Numérique.
Lions, J. L. (1962), *Problèmes aux Limites dans les Équations aux Dérivées Partielles*, Presses de
   l'Université de Montréal.
Lions, J. L. (1968), *Controle Optimal des Systèmes Gouvernés par des Équations aux Dérivées
   Partielles*, Dunod, Paris.
Lions, J. L. (1969), *Quelques Méthodes de Résolution des Problèmes aux Limites non Linéaires*,
   Dunod, Paris.
Lions, J. L. (1977), see Tan Wang Seng and Koh Hock Lee (1977).

Lomax, R. J. (1977), "Preservation of the conservation properties of the finite element method under local mesh refinement," *Comp. Meth. Appl. Mech. Eng.*, **12** (1977), pp. 309–314.

Lyness, J. N. and Jespersen, D. (1975), "Moderate degree symmetric quadrature rules for the triangle," *J. Inst. Math. Applic.*, **15** (1975), pp. 19–32.

El Manouzi, O. H. (1979), "Certaines méthodes d'éléments finis mixtes pour les écoulements visqueux incompressibles," Thèse de 3ème cycle, Université Paris 6.

Manteuffel, T. A. (1978), "The shifted incomplete Cholesky factorization," Report SAND/78/8226, May 1978, Sandia Lab., California.

Manteuffel, T. A. (1979), "Solving structures problems iteratively with a shifted incomplete Cholesky preconditioning," in Glowinski and Lions (1979).

Marrocco, A. (1978a), "Expériences numériques sur des problèmes non linéaires résolus par éléments finis et lagrangien augmenté," Rapport de recherche 309 (LABORIA), IRIA.

Marrocco, A. (1978b), "GEL3D1: module de maillage tridimensionnel automatique 1," Club MODULEF.

Matthies, H. and Strang, G. (1979), "The solution of non linear finite element equations," *Int. J. Num. Met. Eng.*, **14** (1979), pp. 1613–1626.

Meijerink, J. A., and van der Vorst, H. A. (1977), "An iterative solution method for linear systems of which the coefficient matrix is a symmetric M-matrix," *Math. of Comp.*, **31**, (1977), pp. 148–162.

Meijerink, J. A. and van der Vorst, H. A. (1978), "Guide lines for the usage of incomplete decomposition in solving sets of linear equations as occur in practical problems," Technical report TR/9, ACCU, Utrecht, The Netherlands.

Mercier, B. and Pironneau, O. (1977), see Tang Wang Seng and Koh Hock Lee (1977).

Mercier, B. and Tremolieres, R. (1973), "Minimisation de calculs matriciels, applications aux éléments finis et aux problèmes de croisement," Rapport de recherche 3, LABORIA-IRIA.

Miller, J. J. H. (1973, 1977, 1979), "Topics in numerical analysis," *Proceedings of the Royal Irish Academy Conf. on Num. Anal.*, Dublin Publ. by Academic Press.

Mondkar, D. P., and Powell, G. H. (1974), "Towards optimal in-core equation solving," *Computers and Structures*, **4** (1974), pp. 380–392, and 699–728.

Mordant, M. (1979), "Calcul de réacteurs nucléaires en théorie du transport par des méthodes d'éléments finis," in Absi and Glowinski (1979).

Morice, P. (1971), "Calcul parallèle et équations aux dérivées partielles de type elliptique," Technical note IRIA, INF./71016.

Morley, L. S. D. (1968), "The triangular equilibrium element in the solution of plate bending problems," *Aero. Quart.*, **19** (1968), pp. 149–169.

Nallasamy, M. and Krishna Prasad, K. (1977), "On cavity flow at high Reynolds numbers," *J. Fluid Mech.*, **79** (1977), pp. 391–414.

Naves, J. G. (1975), "Calcul d'ensemble voilure-fuselage en régime transonique. Problèmes posés par la mise en exploitation des méthodes d'éléments finis," Proceedings of: 12ème colloque d'Aérodynamique appliquée, 5–7 Nov. 1975, Poitiers, France.

Necas, J. (1967), *Les Méthodes Directes dans les Théories des Équations aux Dérivées Partielles*, Masson, Paris.

Nickell, R. E., Tanner, R. I., and Caswell, B. (1974), "The solution of viscous incompressible jet and free surface flow using finite element method," *J. Fluid Mech.*, **65** (1974), pp. 189–206.

Nitsche, J. A. (1974), "Convergence of non conforming methods," pp. 15–53, in C. de Boor (1974).

Norrie, D. H. (1980), *Proceedings of the Third International Conference on Finite Elements in Flow Problems*, held at Banff, Alberta, Canada, 10–13 June, 1980.

Norrie, D. H. and de Vries, G. (1973), *The Finite Element Method*, Academic Press, London.

Oden, J. T. (1972), "Generalized conjugate functions for mixed finite element approximations of boundary value problems," pp. 629–669, in Aziz (1972).

Oden, J. T. and Lee, J. K. (1977), "Dual mixed hybrid finite element method for second order problems," in Galligani and Magenes (1977), pp. 275–291.

Oden, J. T. and Reddy, J. N. (1976a), *An Introduction to the Mathematical Theory of Finite Elements*, Wiley, New York.

Oden, J. T. and Reddy, J. N. (1976b), *Variational Methods in Theoretical Mechanics*, Springer-Verlag, New York, Heidelberg, Berlin.

Oden, J. T. and Wellford, L. C. (1972), "Analysis of the flow of viscous liquids by the finite element method," *A.I.A.A.*, **10** (1972), pp. 1590–1599.

Oliphant, T. A. (1962). "An extrapolation process for solving linear systems," *Quart. Appl. Math.*, **20** (1962), pp. 257–267.

Olson, M. D. and Shih Yu Tuann (1978), "New finite element results for the square cavity," *Computers and Fluids*, **7** (1978), pp. 123–135.

Paige, C. C. and Saunders, M. A. (1975), "Solution of sparse indefinite systems of linear equations," *SIAM J. Num. Anal.*, **12** (1975), no. 4, pp. 617–629; also: Techn. rep. CS/399, Computer Sc. Dept., Stanford University.

Periaux, J. (1975), "Three dimensional Analysis of compressible potential flows with the finite element method," *Int. J. Numer. Meth. Eng.*, **9**, (1975), pp. 775–831.

Periaux, J. (1978), "Résolution de quelques problèmes non linéaires en aérodynamique par des méthodes d'éléments finis et de moindres carrés," Thèse de 3ème. cycle, Université Paris 6, Laboratoire d'analyse numérique.

Paillard, L. and Benazeth, J. C. (1979), "A "donor-element" method for convective heat transfer analysis," Techn. note MTD/79/125, NOVATOME, Le Plessis Robinson, France.

Perrier, P. and Periaux, P. (1976), "Application of the finite element method in non linear aerodynamics," in *VKI Lecture 87* (Computational Fluid Dynamics, March, 15–19, 1976).

Perronnet, A. (1977a), "Mise en oeuvre informatique de la méthode des éléments finis," Cours de D.E.A., Université Paris 6, Laboratoire d'Analyse Numérique (LA189).

Perronnet, A. (1977b), "The club MODULEF: a library of subroutines for finite element analysis," in Glowinski and Lions (1977).

Pian, T. H. H. (1971), "Formulation of finite element methods for solid continua," pp. 49–83, in Gallagher, Yamada, and Oden (1971).

Pian, T. H. H. (1972), "Finite element formulation by variational principles with relaxed continuity requirements," pp. 557–587, in Aziz (1972).

Pian, T. H. H. and Tong, P. (1969), "Basis of finite elements methods for solid continua," *Int. J. Numer. Meth. Eng.*, **1** (1969), pp. 3–28.

Pironneau, O. (1976), "Sur les problèmes d'optimisation en mécanique des fluides," Thèse de Doctorat d'Etat, Université Paris 6, Mai 1976.

Pironneau, O. (1979), "La méthode des éléments finis et les ordinateurs parallèles SIMD," Rapport de recherche (LABORIA-IRIA) 361, Sept. 1979.

Polak, E. (1971), *Computational Methods in Optimization*, Academic Press, New York and London.

Powell, M. J. D. (1972), "A method for non linear constraints in minimization problems," in *Optimization*, R. Fletcher, ed., Academic Press, New York.

Ramakrishnan, C. V. (1979), "An upwind finite element scheme for the unsteady convective diffusive, transport equation," *Applied Math. Modelling*, **3**, (August, 1979), pp. 280–284.

Rautmann, R. (1980), "Approximation methods for Navier Stokes problems," *Proceedings of IUTAM Symposium*, held at the University of Paderborn, Germany, September 9–15, 1979; Lecture Notes in Mathematics 771, Springer-Verlag.

Raviart, P. A. (1971–1972), "La méthode des éléments finis," Cours du D.E.A. d'analyse numérique, Université Paris 6, Laboratoire d'Analyse Numérique, (LA189).

Raviart, P. A. (1979), "Approximation numérique des phénomènes de diffusion convection. Méthodes d'éléments finis en mécanique des fluides," Cours à l'Ecole d'été d'analyse numérique (EDF/CEA/IRIA).

Raviart, P. A. and Thomas, J. M. (1977a), "A mixed finite element method for second order elliptic problems," pp. 292–315, in Galligani and Magenes (1977).

Raviart, P. A. and Thomas, J. M. (1977b), "Primal hybrid finite element methods for second order elliptic problems," *Math. of Comp.*, **31** (1977), pp. 391–413.

Reid, J. K. ed. (1971), *Large Sparse Sets of Linear Equations*, Academic Press, New York.

Richtmyer, R. D. and Morton, K. W. (1967), *Difference Methods for Initial Value Problems*, Interscience Publishers, Wiley, London.

Roache, P. J. (1972a), *Computational Fluid Dynamics*, Hermosa Publishers, Albuquerque, New Mexico.

Roache, P. J. (1972b), "On artificial viscosity," *J. Comp. Phys.*, **10** (1972), pp. 169.

Roberts, K. V. and Weiss, N. O. (1966), "Convective difference schemes," *Maths. Comp.*, **20** (1966), pp. 272.

Roscoe, D. F. (1974), Ph. D. Thesis, University of Aston, Birmingham.

Roscoe, D. F. (1975), "New methods for the derivation of stable difference representations for differential equations," *J. Inst. Maths. Applic.*, **16** (1975), pp. 291–301.

Roscoe, D. F. (1976), "The solution of the three dimensional Navier-Stokes equations using a new finite difference approach," *Int. J. Num. Meth. Eng.*, **10** (1976), pp. 1299–1308.

Rose, D. J. and Willoughby, eds. (1972), *Sparse Matrices and Their Applications*, Plenum Press, London.

Saylor, P. (1974), "Second order strongly implicit symmetric factorization methods for the solution of elliptic difference equations," *SIAM J. Numer. Anal.*, **11** (1974), pp. 894–908.

Segal, A. (1977), "On the numerical solution of Stokes equation using the finite element method," Report NA/16(1977), Math. Inst. Techn. Hogeschool, Delft, Nederland; *Comp. Meth. Appl. Mech. Eng.*, **19** (1979), pp. 165–185.

Shestakov, A. I. (1979), "A hybrid vortex-ADI solution for flows of low viscosity," *J. Comp. Ph.*, **31** (1979), pp. 313–334.

Spalding, D. B. (1972), *Int. J. Num. Meth. Eng.*, **4** (1972), p. 551.

Stone, H. L. (1968), "Iterative solution of implicit approximations of multidimensional partial differential equations," *SIAM J. Numer. Anal.*, **5** (1968), pp. 530–558.

Strang, G. (1972a), "Approximation in the finite element method," *Numerische Mathematik*, **19** (1972), pp. 81–98.

Strang, G. (1972b), "Variational crimes in the finite element method," pp. 689–710 in Aziz (1972).

Strang, G. and Fix, G. J. (1972), *An Analysis of the Finite element Method*, Prentice-Hall, Englewood Cliffs, New Jersey.

Stroud, A. H. (1971), *Approximate Calculation of Multiple Integrals*, Prentice Hall, Englewood Cliffs, New Jersey.

Stummel, F. (1980), "The limitations of the patch test," *Int. J. Num. Meth. Eng.*, **15** (1980), pp. 177–188.

Synge, J. L. (1957), *The Hypercircle in Mathematical Physics*, Cambridge University Press.

Tabata, M. (1977), "A finite element approximation corresponding to the upwind differencing," *Memoirs of Numerical Mathematics*, **1** (1977), pp. 47–63.

Tanner, R. I., Nickell, R. E., and Bilger, R. W. (1975), "Finite element methods for the solution of some incompressible non newtonian fluid mechanics problems with free surfaces," *Comp. Meth. Appl. Mech. Eng.*, **6** (1975), pp. 155–174.

Tan Wang Seng and Koh Hock Lee (1977), *Lectures on the Finite Element Methods*; part 1: "Topics in numerical analysis," by J. L. Lions; part 2: "Some examples of implementation and application of the finite element method," by B. Mercier and O. Pironneau; University of Malaysia Press, Penang, Malaysia; (The 2nd. part is also edited as rapport de recherche 248, IRIA-LABORIA).

Taylor, C. and Hood, P. (1973), "A numerical solution of the Navier-Stokes equations using the finite element technique," *Computers and Fluids*, **1** (1973), pp. 73–100.

Temam, R. (1968), "Une méthode d'approximation des équations de Navier-Stokes," *Bulletin de la Société Mathématique de France*, **98** (1968), pp. 115–152.

Temam, R. (1975), "On the Euler equations of incompressible perfect fluids," *J. of Functional Analysis*, **20** (1975), pp. 32–43.

Temam, R. (1977a), *Theory and Numerical Analysis of the Navier-Stokes Equations*, North-Holland, Amsterdam.

Temam, R. (1977b), "Some finite element methods in fluid flow," *VKI Lecture Series*, no. 86, (Computational fluid dynamics, March 21–25, 1977).

Temam, R. and Thomasset, F. (1976), "Numerical solution of Navier-Stokes equations by a finite element method," in Gallagher, Morandi-Cecchi *et. al.* (1976).

Tewarson, R. P. (1970), "Computation with sparse matrices," *SIAM Review*, **12** (1970), no. 4, pp. 527–543.

Tewarson, R. P. (1973), *Sparse Matrices*, Academic Press, New York.

Thacker, W. C., Gonzalez, A., and Putland, G. E. (1977), "A method for automating the construction of irregular computational grids for storm sturge forecast models," preprint; Sea-air Interaction Lab., Miami, Florida.

Thomas, J. M. (1976), "Méthode des éléments finis hybrides duaux pour les problèmes elliptiques du second ordre," *R.A.I.R.O., Série Analyse Numérique*, **10** (1976), pp. 51–79.

Thomas, J. M. (1977), "Sur l'analyse numérique des méthodes d'éléments finis hybrides et mixtes," Université Paris 6, Laboratoire d'analyse numérique.

Thomasset, F. (1977), "Numerical solution of the Navier-Stokes equations by finite elements methods," *VKI Lecture Series*, no. 86, (Computational Fluid Dynamics, March 21–25, 1977).

Thomasset, F. (1980), "Manuel d'utilisation des modules NSNCST and NSNCEV (Club MODULEF)" INRIA, Rocquencourt, France.

de Vahl Davis, G. and Mallinson, G. D. (1973), "The method of false transient for the solution of coupled elliptic equations," *J. Comp. Ph.*, **12** (1973), pp. 435–461.

de Vahl Davis, G. and Mallinson, G. D. (1976), "An evaluation of upwind and central difference approximations by a study of recirculating flow," *Computers and Fluids*, **4** (1976), pp. 29–43.

van-Ingelandt, M. (1979), "Modules de résolution des systèmes linéaires stockés sous forme hypermatricielle sur mémoire secondaire à accès direct," rapport no. 149 (1979), Université des sciences and techniques de Lille, France; also pub. of club Modulef, see Begis and Perronnet (1979).

Varga, R. S. (1961), *Matrix Iterative Analysis*, Prentice-Hall, Englewood, New Jersey.

Vendham, C. P., Kapoor, M. P., and Das, Y. C. (1974), "An integrated sequential solver for large matrix equations," *Int. J. Num. Meth. Eng.*, **8** (1974), pp. 227–248.

Viaud, D. (1970), "Approximation des équations de Navier-Stokes par des méthodes de pénalisation," Thèse de 3ème cycle, Université Paris 6.

Vidrascu, M. (1978), "Sur la résolution numérique du problème de Dirichlet pour l'opérateur biharmonique," thèse de 3ème cycle, Université Paris 6, Laboratoire d'Analyse Numérique.

Wilson, E. L., Bathe, K. J., and Doherty, W. P. (1974), "Direct solution of large systems of linear equations," *Computers and Structures*, **4** (1974), pp. 363–372.

Young, D. M. (1971), *Iterative Solution of Large Linear Systems*, Academic Press, New York.

Zienckiewicz, O. C. (1977), *The Finite Element Method in Engineering Science*, McGraw-Hill, New York.

Zienckiewicz, O. C. and Godbole, P. N. (1975), Viscous incompressible flow with special reference to non newtonian fluids," in Gallagher *et. al.* (1975).

Zlamal, M. (1968), "On the finite element method," *Numerische Mathematik*, **12** (1968), pp. 394–409.

Zlamal, M. (1973), "The finite element method in domains with curved boundaries," *Int. J. Numer. Meth. Eng.*, **5** (1973), pp. 367–373.

Zlamal, M. (1974), "Curved elements in the finite element method," *SIAM J. Numer. Anal.*, **10** (1973), pp. 229–240, and 11 (1974), pp. 347–362.

# Index

area coordinates 18

bandwidth 55, 58, 128
barycentric coordinates 18
bubble function 79, 82

centered differences 37
characteristics 55
Cholesky factorisation 128
curved finite elements 15, 20

discontinuous finite elements 50
diffusion-advection equation 38, 44,
    48, 57
divergence free basis 77, 85, 142
downwinding 39

fill-in 128
frontal method 86, 131

Hermite finite elements 15

incompressible flow 1, 56, 72
isoparametric flow 15, 20, 22

Lagrange finite elements 15
Lagrange multipliers 73, 76

mass-lumping 48, 71
maximum principle 49
mixed finite elements 27
M-matrix 20, 27
MODULEF 2, 120, 133

nonconforming finite elements 25, 67,
    76, 103, 136
numerical integration 22
numerical viscosity 39

patch test 26, 136

reference element 21, 23

saddle points 36
SIMD architecture 2
singularity points 17
solenoidal basis 77, 85, 142
square cavity 107

upstream information 38, 51

wiggles 38, 60

zero divergence basis 77, 85, 142

# Springer Series
# in Computational Physics

Editors: W. Beiglböck, H. Cabannes, H. B. Keller, J. Killeen, S. A. Orszag

**Numerical Methods in Fluid Dynamics**
**M. Holt,** University of California-Berkeley
1977. viii, 253 pages. 107 illustrations. 2 tables. cloth.
ISBN 0-387-**07907**-6

**A Computational Method in Plasma Physics**
**P. Garabedian, F. Bauer,** and **O. Betancourt,** New York
University
1978 vi, 144 pages. 22 figures. cloth.
ISBN 0-387-**08833**-4

**Unsteady Viscous Flows**
**D. Telionis,** Virginia Polytechnic Institute and State
University College of Engineering
1981. approx. 406 pages. 127 figures. cloth.
ISBN 0-387-**10481**-X

**Finite-Difference Techniques for Vectorized**
**Fluid Dynamics Calculations**
**D. Book,** Naval Research Laboratory
1981. 240 pages. 60 figures. cloth.
ISBN 0-387-**10482**-8

**Implementation of Finite Element Methods**
**for Navier-Stokes Equations**
**F. Thomasset,** Institut National De Recherche En Informatique
Et En Automatique, Domaine de Voluceau,
Rocquencourt, F-78150 Le Chesnay, France
1981. 176 pages. 86 figures. cloth.
ISBN 0-387-**10771**-1